P9-ELX-632

50 physics ideas

you really need to know

Joanne Baker

Quercus

Contents

Introduction

When I told my friends about this book they joked that the first thing you really need to know about physics is that it is hard. Despite this, we all use physics every day. When we look in a mirror, or put on a pair of glasses we are using the physics of optics. When we set our alarm clocks we track time; when we follow a map we navigate geometric space. Our mobile phones connect us via invisible electromagnetic threads to satellites orbiting overhead. But physics is not all about technology. Without it there would be no moon, no rainbows, no diamonds. Even the blood flowing through our arteries follows the laws of physics, the science of the physical world.

Modern physics is full of surprises. Quantum physics turned our world upside down by questioning the very concept of an object's existence. Cosmology asks what the universe is. How did it arise and why are we here? Is our universe special or somehow inevitable? Peering inside atoms, physicists uncovered a hidden ghostly world of fundamental particles. Even the most solid mahogany table is mostly made of empty space, its atoms supported by a scaffolding of nuclear forces. Physics grew out of philosophy, and in a way it is turning back towards it by providing new and unexpected views of the world that surpass our daily experiences.

However, physics is not just a collection of imaginative ideas. It is rooted in fact and experiment. The scientific method continually upgrades the laws of physics, like computer software, as bugs are fixed and new modules added. If the evidence requires it, major shifts in thinking can be accommodated, but acceptance takes time. It took more than a generation for Copernicus's idea that the Earth goes round the Sun to be widely accepted, but the pace has quickened and quantum physics and relativity were integrated into physics within a decade. So even the most successful laws of physics are constantly being tested.

This book offers you a whistle-stop tour of the world of physics, from basic concepts like gravity, light and energy through to modern ideas of quantum theory, chaos and dark energy. I hope that, like a good tourist guidebook, it tempts you to find out more. Physics is not just fundamental – it is *fun*.

01 Mach's principle

A child whirling on a merry-go-round is tugged outwards by the distant stars. This is Mach's principle that 'mass there influences inertia here'. Through gravity, objects far away affect how things move, and spin, nearby. But why is this and how can you tell if something is moving or not?

If you have ever sat in a train at a station and seen through the window a neighbouring carriage pull away from yours, you will know that sometimes it is hard to tell whether it is your own train leaving the station or the other arriving. Is there a way that you could measure for sure which one is in motion?

Ernst Mach, Austrian philosopher and physicist, grappled with this question in the 19th century. He was treading in the footsteps of the great Isaac Newton who had believed, unlike Mach, that space was an absolute backdrop. Like graph paper, Newton's space contained an engraved set of coordinates and he mapped all motions as movements with respect to that grid. Mach, however, disagreed, arguing instead that motion was only meaningful if measured with respect to another object, not the grid. What does it mean to be moving if not relative to something else? In this sense Mach, who was influenced by the earlier ideas of Newton's competitor Gottfried Leibniz, was a forerunner to Albert Einstein in preferring to think that only relative motions made sense. Mach argued that because a ball rolls in the same way whether it is in France or Australia, the grid of space is irrelevant. The only thing that can conceivably affect how the ball rolls is gravity. On the Moon the ball might well roll differently because

timeline

c.335BC
Aristotle states that objects move due to the action of forces

AD1640
Galileo formulates the principle of inertia

> **'Absolute space, of its own nature without reference to anything external, always remains homogenous and immovable.'**
>
> **Isaac Newton, 1687**

the gravitational force pulling on the ball's mass is weaker there. Because every object in the universe exerts a gravitational pull on every other, each object will feel each other's presence through their mutual attractions. So motion must ultimately depend on the distribution of matter, or its mass, not on the properties of space itself.

Mass What exactly is mass? It is a measure of how much matter an object contains. The mass of a lump of metal would be equal to the sum of the masses of all the atoms in it. Mass is subtly different from weight. Weight is a measure of the force of gravity pulling a mass down – an astronaut weighs less on the Moon than on Earth because the gravitational force exerted by the smaller Moon is less. But the astronaut's mass is the same – the number of atoms he contains has not changed. According to Albert Einstein, who showed that energy and mass are interchangeable, mass can be converted into pure energy. So mass is, ultimately, energy.

Inertia Inertia, named after the Latin word for 'laziness', is very similar to mass but tells us how hard it is to move something by applying a force. An object with large inertia resists movement. Even in outer space a massive object takes a large force to move it. A giant rocky asteroid on collision course with the Earth may need a huge shove to deflect it, whether it is created by a nuclear explosion or a smaller force applied for a longer time. A smaller spacecraft, with less inertia than the asteroid, might be manoeuvred easily with tiny jet engines.

The Italian astronomer Galileo Galilei proposed the principle of inertia in the 17th century: if an object is left alone, and no forces are applied to it,

1687
Newton publishes his bucket argument

1893
Mach publishes *The Science of Mechanics*

1905
Einstein publishes the special theory of relativity

then its state of motion is unchanged. If it is moving, it continues to move at the same speed and in the same direction. If it is standing still it continues to do so. Newton refined this idea to form his first law of motion.

Newton's bucket Newton also codified gravity. He saw that masses attract one another. An apple falls from a tree to the ground because it is attracted by the Earth's mass. Equally, the Earth is attracted by the apple's mass, but we would be hard pressed to measure the microscopic shift of the whole Earth towards the apple.

Newton proved that the strength of gravity falls off quickly with distance, so the Earth's gravitational force is much weaker if we are floating high above it rather than on its surface. But nevertheless we would still feel the reduced pull of the Earth. The further away we go the weaker it would get, but it could still tweak our motion. In fact, all objects in the universe may exert a tiny gravitational pull that might subtly affect our movement.

Newton tried to understand the relationships between objects and movement by thinking about a spinning bucket of water. At first when the bucket is turned, the water stays still even though the bucket moves. Then the water starts to spin as well. Its surface dips as the liquid tries to escape by creeping up the sides but it is kept in place by the bucket's confining force. Newton argued that the water's rotation could only be understood if seen in the fixed reference frame of absolute space, against its grid. We could tell if the bucket was spinning just by looking at it because we would see the forces at play on it producing the concave surface of the water.

Centuries later Mach revisited the argument. What if the water-filled bucket were the only thing in the universe? How could you know it was the bucket that was rotating? Couldn't you equally well say the water was rotating relative to the bucket? The only way to make sense of it would be to place another object into the bucket's universe, say the wall of a room, or even a distant star. Then the bucket would clearly be spinning relative to that. But without the frame of a stationary room, and the fixed stars, who could say whether it was the bucket or the water that rotates? We experience the same thing when we watch the Sun and stars arc across the sky. Is it the stars or the Earth that is rotating? How can we know?

ERNST MACH 1838–1916

As well as for Mach's principle, Austrian physicist Ernst Mach is remembered for his work in optics and acoustics, the physiology of sensory perception, the philosophy of science and particularly his research on supersonic speed. He published an influential paper in 1877 that described how a projectile moving faster than the speed of sound produces a shock wave, similar to a wake. It is this shockwave in air that causes the sonic boom of supersonic aircraft. The ratio of the speed of the projectile, or jet plane, to the speed of sound is now called the Mach number, such that Mach 2 is twice the speed of sound.

According to Mach, and Leibniz, motion requires external reference objects for us to make sense of it, and therefore inertia as a concept is meaningless in a universe with just one object in it. So if the universe were devoid of any stars, we'd never know that the Earth was spinning. The stars tell us we're rotating relative to them.

The ideas of relative versus absolute motion expressed in Mach's principle have inspired many physicists since, notably Einstein (who actually coined the name 'Mach's principle'). Einstein took the idea that all motion is relative to build his theories of special and general relativity. He also solved one of the outstanding problems with Mach's ideas: rotation and acceleration must create extra forces, but where were they? Einstein showed that if everything in the universe were rotating relative to the Earth, we should indeed experience a small force that would cause the planet to wobble in a certain way.

The nature of space has puzzled scientists for millennia. Modern particle physicists think it is a seething cauldron of subatomic particles being continually created and destroyed. Mass, inertia, forces and motion may all in the end be manifestations of a bubbling quantum soup.

the condensed idea
Mass matters for motion

02 Newton's laws of motion

Isaac Newton was one of the most prominent, contentious and influential scientists of all time. He helped to invent calculus, explained gravity and identified the constituent colours of white light. His three laws of motion describe why a golf ball follows a curving path, why we are pressed against the side of a cornering car and why we feel the force through a baseball bat as it strikes the ball.

Although motorcycles had yet to be invented in Newton's time, his three laws of motion explain how a stunt rider can mount the vertical wall of death, and how Olympic cyclists race on inclined tracks.

Newton, who lived in the 17th century, is considered one of the foremost intellects of science. It took his highly inquisitive character to understand some of the most seemingly simple yet profound aspects of our world, such as how a thrown ball curves through the air, why things fall down rather than up and how the planets move around the Sun.

An average student at Cambridge in the 1660s, Newton began by reading the great works of mathematics. Through them he was drawn away from civic law into the laws of physics. Then, on sabbatical at home when the university was closed for an outbreak of plague, Newton took the first steps to developing his three laws of motion.

Forces Borrowing Galileo's principle of inertia, Newton formulated his first law. It states that bodies do not move or change their speed unless a force acts. Bodies that are not moving will remain stationary unless a force

timeline

c.350BC
Aristotle proposes in *Physics* that motions are due to ongoing changes

AD1640
Galileo formulates the principle of inertia

Newton's laws of motion

First law Bodies move in a straight line with a uniform speed, or remain stationary, unless a force acts to change their speed or direction

Second law Forces produce accelerations that are in proportion to the mass of a body ($F = ma$)

Third law Every action of a force produces an equal and opposite reaction

is applied; bodies that are moving with some constant speed keep moving at that same speed unless acted upon by a force. A force (for instance a push) supplies an acceleration that changes the velocity of the object. Acceleration is a change in speed over some time.

This is hard to appreciate in our own experience. If we throw a hockey puck it skims along the ice but eventually slows due to friction with the ice. Friction causes a force that decelerates the puck. But Newton's first law may be seen in a special case where there is no friction. The nearest we might get to this is in space, but even here there are forces such as gravity at work. Nevertheless, this first law provides a basic touchstone from which to understand forces and motion.

Acceleration Newton's second law of motion relates the size of the force to the acceleration it produces. The force needed to accelerate an object is proportional to the object's mass. Heavy objects – or rather ones with large inertia – need more force to accelerate them than lighter objects. So to accelerate a car from standing still to 100 kilometres an hour in one minute would take a force equal to the car's mass times its increase in speed per unit time.

1687
Newton publishes the *Principia*

1905
Einstein publishes the special theory of relativity

Newton's second law is expressed algebraically as '$F = ma$', force (F) equals mass (m) times acceleration (a). Turning this definition around, the second law expressed in another way says that acceleration is equal to force per unit mass. For a constant acceleration, force per unit mass is also unchanged. So the same amount of force is needed to move a kilogram mass whether it is part of a small or large body. This explains Galileo's imaginary experiment that asks which would hit the ground first if dropped together: a cannonball or a feather? Visualizing it we may think that the cannonball would arrive ahead of the drifting feather. But this is simply due to the air resistance that wafts the feather. If there were no air, then both would fall at the same rate, hitting the ground together. They experience the same acceleration, gravity, so they fall side by side. *Apollo 15* astronauts showed in 1971 that on the Moon, where there is no atmosphere to slow it down, the feather falls at the same rate as a geologist's heavy hammer.

Action equals reaction Newton's third law states that any force applied to a body produces an equal and opposite reaction force in that body. In other words, for every action there is a reaction. The opposing force is felt as recoil. If one roller-skater pushes another, then she will also roll backwards as she pushes against her partner's body. A marksman feels the kick of the rifle against his shoulder as he shoots. The recoil force is equal in size to that originally expressed in the shove or the bullet. In crime films the victim of a shooting often gets propelled backwards by the force of the bullet. This is misleading. If the force was really so great then the shooter should also be hurled back by the recoil of his gun. Even if we jump up off the ground, we exert a small downward force on the Earth, but because the Earth is so much more massive than we are, it barely shows.

With these three laws, plus gravity, Newton could explain the motion of practically all objects, from falling acorns to balls fired from a cannon. Armed with these three equations he could confidently have climbed aboard a fast motorbike and sped up onto the wall of death, had such a thing existed in his day. How much trust would you place in Newton's laws? The first law says that the cycle and its rider want to keep travelling in one direction at a certain speed. But to keep the cycle moving in a circle, according to the second law, a confining force needs to be provided to continually change its direction, in this case applied by the track through the wheels. The force needed is equal to the mass of the cycle and

ISAAC NEWTON 1643–1727

Isaac Newton was the first scientist to be honoured with a knighthood in Britain. Despite being 'idle' and 'inattentive' at school, and an unremarkable student at Cambridge University, Newton flourished suddenly when plague closed the university in the summer of 1665. Returning to his home in Lincolnshire, Newton devoted himself to mathematics, physics and astronomy, and even laid the foundations for calculus. There he produced early versions of his three laws of motion and deduced the inverse square law of gravity. After this remarkable outburst of ideas, Newton was elected to the Lucasian Chair of Mathematics in 1669 at just 27 years old. Turning his attention to optics, Newton discovered with a prism that white light was made up of rainbow colours, quarrelling famously with Robert Hooke and Christiaan Huygens over the matter. Newton wrote two major works, *Philosophiae naturalis Principia mathematica*, or *Principia*, and *Opticks*. Late in his career, Newton became politically active. He defended academic freedom when King James II tried to interfere in university appointments and entered Parliament in 1689. A contrary character, on the one hand desiring attention and on the other being withdrawn and trying to avoid criticism, Newton used his powerful position to fight bitterly against his scientific enemies and remained a contentious figure until his death.

rider multiplied by their acceleration. The third law then explains the pressure exerted by the cycle on the track, as a reactionary force is set up. It is this pressure that glues the stunt rider to the inclined wall, and if the bike goes fast enough it can even ride on a vertical wall.

Even today knowledge of Newton's laws is pretty much all you need to describe the forces involved in driving a car fast around a bend or, heaven forbid, crashing it. Where Newton's laws do not hold is for things moving close to the speed of light or with very small masses. It is in these extremes that Einstein's relativity and the science of quantum mechanics take over.

the condensed idea
Motion captured

03 Kepler's laws

Johannes Kepler looked for patterns in everything. Peering at astronomical tables describing the looped motions of Mars projected on the sky, he discovered three laws that govern the orbits of the planets. Kepler described how planets follow elliptical orbits and how more distant planets orbit more slowly around the Sun. As well as transforming astronomy, Kepler's laws laid the foundations for Newton's law of gravity.

As the planets move around the Sun, the closest ones move more quickly around it than those further away. Mercury circles the Sun in just 80 Earth days. If Jupiter travelled at the same speed it would take about 3.5 Earth years to complete an orbit when, in fact, it takes 12. As all the planets sweep past each other, when viewed from the Earth some appear to backtrack as the Earth moves forwards past them. In Kepler's time these 'retrograde' motions were a major puzzle. It was solving this puzzle that gave Kepler the insight to develop his three laws of planetary motion.

'It suddenly struck me that that tiny pea, pretty and blue, was the Earth. I put up my thumb and shut one eye, and my thumb blotted out the planet Earth. I didn't feel like a giant. I felt very, very small.'

Neil Armstrong, b.1930

Patterns of polygons The German mathematician Johannes Kepler sought patterns in nature. He lived in the late 16th and early 17th centuries, a time when astrology was taken very seriously and astronomy as a physical science was still in its infancy. Religious and spiritual ideas were just as important in revealing nature's laws as observation. A mystic who believed that the underlying structure of the universe was built from perfect geometric forms, Kepler devoted his life to trying to tease out the patterns of imagined perfect polygons hidden in nature's works.

timeline

c.580 BC
Pythagoras states planets orbit in perfect crystalline spheres

c.AD 150
Ptolemy records retrograde motion and suggests planets move in epicycles

JOHANNES KEPLER 1571–1630

Johannes Kepler liked astronomy from an early age, recording in his diary a comet and a lunar eclipse before he was ten. While teaching at Graz, Kepler developed a theory of cosmology that was published in the *Mysterium Cosmographicum* (*The Sacred Mystery of the Cosmos*). He later assisted astronomer Tycho Brahe at his observatory outside Prague, inheriting his position as Imperial Mathematician in 1601. There Kepler prepared horoscopes for the emperor and analysed Tycho's astronomical tables, publishing his theories of noncircular orbits, and the first and second laws of planetary motion, in *Astronomia Nova* (*New Astronomy*). In 1620, Kepler's mother, a herbal healer, was imprisoned as a witch and only released through Kepler's legal efforts. However, he managed to continue his work and the third law of planetary motion was published in *Harmonices Mundi* (*Harmony of the Worlds*).

Kepler's work came a century after Polish astronomer Nicolaus Copernicus proposed that the Sun lies at the centre of the universe and the Earth orbits the Sun, rather than the other way around. Before then, going back to the Greek philosopher Ptolemy, it was believed that the Sun and stars orbited the Earth, carried on solid crystal spheres. Copernicus dared not publish his radical idea during his lifetime, leaving it to his colleague to do so just before he died, for fear that it would clash with the doctrine of the church. Nevertheless Copernicus caused a stir by suggesting that the Earth was not the centre of the universe, implying that humans were not the most important beings in it, as favoured by an anthropocentric god.

Kepler adopted Copernicus's heliocentric idea, but nevertheless still believed that the planets orbited the Sun in circular orbits. He envisaged a system in which the planets' orbits lay within a series of nested spheres spaced according to mathematical ratios that were derived from the sizes of

1543
Copernicus proposes planets orbit the Sun

1576
Tycho Brahe maps the planets' positions

1609
Kepler discovers that planets move in elliptical orbits

1687
Newton explains Kepler's laws with his law of gravitation

'We are just an advanced breed of monkeys on a minor planet of a very average star. But we can understand the universe. That makes us something very special.'

Stephen Hawking, 1989

three-dimensional shapes that would fit within them. So he imagined a series of polygons with increasing numbers of sides that fit within the spheres. The idea that nature's laws followed basic geometric ratios had originated with the Ancient Greeks.

The word planet comes from the Greek for 'wanderer'. Because the other planets in our solar system lie much closer to the Earth than the distant stars, they appear to wander across the sky. Night after night they pick out a path through the stars. Every now and again, however, their path reverses and they make a little backwards loop. These retrograde motions were thought to be bad omens. In the Ptolemaic model of planetary motion this behaviour was impossible to understand, and so astronomers added 'epicycles' or extra loops to the orbit of a planet that mimicked this motion. But the epicycles did not work very well. Copernicus's Sun-centred universe needed fewer epicycles than the older Earth-centred one, but still could not explain the fine details.

Kepler's laws

First law planetary orbits are elliptical with the Sun at one focus

Second law a planet sweeps out equal areas in equal times as it orbits the Sun

Third law the orbital periods scale with ellipse size, such that the period squared is proportional to the semi-major axis length cubed

Trying to model the orbits of the planets to support his geometric ideas, Kepler used the most accurate data available, intricate tables of the planets' motions on the sky, painstakingly prepared by Tycho Brahe. In these columns of numbers Kepler saw patterns that suggested his three laws.

Kepler got his breakthrough by disentangling the retrograde motions of Mars. He recognized that the backward loops would fit if the planets' orbits were elliptical around the Sun and not circular as had been thought. Ironically this meant that nature did not follow perfect shapes. Kepler must have been overjoyed at his success in fitting the orbits, but also shocked that his entire philosophy of pure geometry had been proved wrong.

Orbits In Kepler's first law, he noted that the planets move in elliptical orbits with the Sun at one of the two foci of the ellipse.

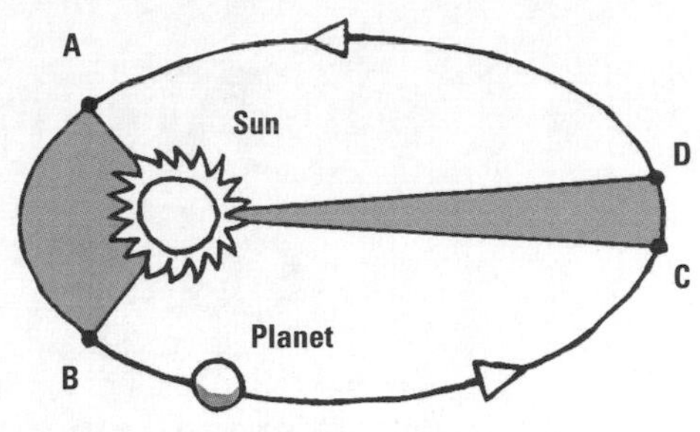

Kepler's second law describes how quickly a planet moves around its orbit. As the planet moves along its path, it sweeps out an equal area segment in an equal time. The segment is measured using the angle drawn between the Sun and the planet's two positions (*AB* or *CD*), like a slice of pie. Because the orbits are elliptical, when the planet is close to the Sun it needs to cover a larger distance to sweep out the same area than when it is further away. So the planet moves faster near the Sun than when it is distant. Kepler's second law ties its speed with its distance from the Sun. Although Kepler didn't realize it at the time, this behaviour is ultimately due to gravity accelerating the planet faster when it is near the Sun's mass.

Kepler's third law goes one step further again and tells us how the orbital periods scale up for different sized ellipses at a range of distances from the Sun. It states that the squares of the orbital periods are proportional to the cube power of the longest axis of the elliptical orbit. The larger the elliptical orbit, the slower the period, or time taken to complete an orbit. So planets further from the Sun orbit more slowly than nearby planets. Mars takes nearly 2 Earth years to go around the Sun, Saturn 29 years and Neptune 165 years.

> **'I measured the skies, now the shadows I measure,**
> **Sky-bound was the mind, earth-bound the body rests.'**
> Kepler's epitaph, 1630

In these three laws, Kepler managed to describe all the planets' orbits in our solar system. His laws apply equally to any body in orbit around another, from comets, asteroids and moons in our solar system to planets around other stars and even artificial satellites whizzing around the Earth. Kepler succeeded in unifying the principles into geometric laws but he did not know why these laws held. He believed that they arose from the underlying geometric patterns of nature. It took Newton to unify these laws into a universal theory of gravity.

the condensed idea
Law of the worlds

04 Newton's law of gravitation

Isaac Newton made a giant leap when he connected the motions of cannonballs and fruit falling from trees to the movements of the planets, thus linking heaven and earth. His law of gravitation remains one of the most powerful ideas of physics, explaining much of the physical behaviour of our world. Newton argued that all bodies attract each other through the force of gravity and the strength of that force drops off with distance squared.

The idea of gravity supposedly came to Isaac Newton when he saw an apple fall from a tree. We don't know if this is true or not, but Newton stretched his imagination from earthly to heavenly motions to work out his law of gravitation.

'Gravity is a habit that is hard to shake off.'
Terry Pratchett, 1992

Newton perceived that objects were attracted to the ground by some accelerating force (see page 8). If apples fall from trees, what if the tree were even higher? What if it reached to the Moon? Why doesn't the Moon fall to the Earth like an apple?

All fall down Newton's answer lay first in his laws of motion linking forces, mass and acceleration. A ball blasted from a cannon travels a certain distance before falling to the ground. What if it were fired more quickly? Then it would travel further. If it was fired so fast that it travelled far enough in a straight line that the Earth curved away beneath it, where

timeline

350BC
Aristotle discusses why objects fall

AD1609
Kepler reveals the laws of planetary orbits

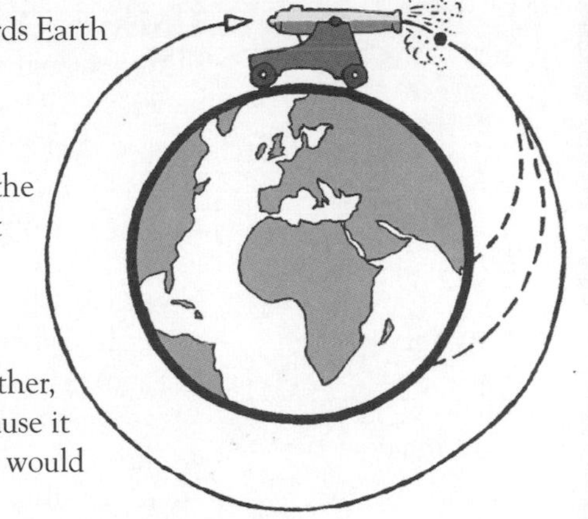

would it fall? Newton realized that it would be pulled towards Earth but would then follow a circular orbit. Just like a satellite constantly being pulled but never reaching the ground.

When Olympic hammer-throwers spin on their heels, it is the pull on the string that keeps the hammer rotating. Without this pull the hammer would fly off in a straight line, just as it does on its release. It's just the same with Newton's cannonball – without the centrally directed force tying the projectile to Earth, it would fly off into space. Thinking further, Newton reasoned that the Moon also hangs in the sky because it is held by the invisible tie of gravity. Without gravity it too would fly off into space.

Inverse square law Newton then tried to quantify his predictions. After exchanging letters with Robert Hooke, Newton showed that gravity follows an inverse square law – the strength of gravity decreases by the square of the distance from a body. So if you travel twice some distance from a body its gravity is four times less; the gravity exerted by the Sun would be four times less for a planet in an orbit twice as far from it as the Earth, or a planet three times distant would experience gravity nine times less.

Newton's inverse square law of gravity explained in one equation the orbits of all the planets as described in the three laws of Johannes Kepler (see page 12). Newton's law predicted that they travelled quicker near the Sun as they followed their elliptical paths. A planet feels a stronger gravitational force from the Sun when it travels close to it, which makes it speed up. As its speed increases the planet is thrown away from the Sun again, gradually slowing back down. Thus, Newton pulled together all the earlier work into one profound theory.

> **‘Every object in the universe attracts every other object along a line of the centres of the objects, proportional to each object's mass, and inversely proportional to the square of the distance between the objects.’**
>
> **Isaac Newton,** 1687

1640	1687	1905	1915
Galileo states the principle of inertia	Newton's *Principia* is published	Einstein publishes the special theory of relativity	Einstein publishes the general theory of relativity

Universal law Generalizing boldly, Newton then proposed that his theory of gravity applied to everything in the universe. Any body exerts a gravitational force in proportion to its mass, and that force falls off as the inverse square of distance from it. So any two bodies attract each other. But because gravity is a weak force we only really observe this for very massive bodies, such as the Sun, Earth and planets.

On the surface of the Earth the acceleration due to gravity, *g*, is 9.8 metres per second per second.

If we look closer, though, it is possible to see tiny variations in the local strength of gravity on the surface of the Earth. Because massive mountains and rocks of differing density can raise or reduce the strength of gravity near them, it is possible to use a gravity meter to map out geographic terrains and to learn about the structure of the Earth's crust. Archaeologists also sometimes use tiny gravity changes to spot buried settlements. Recently, scientists have used gravity-measuring space satellites to record the (decreasing) amount of ice covering the Earth's poles and also to detect changes in the Earth's crust following large earthquakes.

Back in the 17th century, Newton poured all his ideas on gravitation into one book, *Philosophiae naturalis principia mathematica*, known as the *Principia*. Published in 1687, the *Principia* is still revered as a scientific milestone. Newton's universal gravity explained the motions not only of planets and moons but also of projectiles, pendulums and apples. He

The discovery of Neptune

The planet Neptune was discovered thanks to Newton's law of gravitation. In the early 19th century, astronomers noticed that Uranus did not follow a simple orbit but acted as though another body was disturbing it. Various predictions were made based on Newton's law and in 1846 the new planet, named Neptune after the sea god, was discovered close to the expected position. British and French astronomers disagreed over who had made the discovery, which is credited to both John Couch Adams and Urbain Le Verrier. Neptune has a mass 17 times that of the Earth and is a 'gas giant' with a thick dense atmosphere of hydrogen, helium, ammonia and methane smothering a solid core. The blue colour of Neptune's clouds is due to methane. Its winds are the strongest in the solar system, reaching as much as 2500 kilometres per hour.

Tides

Newton described the formation of ocean tides on the Earth in his book the *Principia*. Tides occur because the Moon pulls differently on oceans on the near and far sides of the Earth, compared with the solid Earth itself. The different gravitational pull on opposite sides of the Earth causes the surface water to bulge both towards and away from the Moon, leading to tides that rise and fall every 12 hours. Although the more massive Sun exerts a stronger gravitational force on the Earth than the smaller Moon, the Moon has a stronger tidal effect because it it closer to the Earth. The inverse square law means that the gravitational gradient (the difference felt by the near and far sides of the Earth) is much greater for the closer Moon than the distant Sun. During a full or new Moon, the Earth, Sun and Moon are all aligned and especially high tides result, called 'spring' tides. When these bodies are misaligned, at 90 degrees to one another, weak tides result called 'neap' tides.

explained the orbits of comets, the formation of tides and the wobbling of the Earth's axis. This work cemented Newton's reputation as one of the great scientists of all time.

'It has been said that arguing against globalization is like arguing against the laws of gravity.'

Kofi Annan, b.1938

Newton's universal law of gravitation has stood for hundreds of years and still today gives a basic description of the motion of bodies. However, science does not stand still, and 20th-century scientists built upon its foundations, notably Einstein with his theory of general relativity. Newtonian gravity still works well for most objects we see and for the behaviour of planets, comets and asteroids in the solar system that are spread over large distances from the Sun where gravity is relatively weak. Although Newton's law of gravitation was powerful enough to predict the position of the planet Neptune, discovered in 1846 at the expected location beyond Uranus, it was the orbit of another planet, Mercury, that required physics beyond that of Newton. Thus general relativity is needed to explain situations where gravity is very strong, such as close to the Sun, stars and black holes.

the condensed idea
Mass attraction

05 Conservation of energy

Energy is an animating force that makes things move or change. It comes in many guises and may manifest itself as a change in height or speed, travelling electromagnetic waves or the vibrations of atoms that cause heat. Although energy can metamorphose between these types, the overall amount of energy is always conserved. More cannot be created and it can never be destroyed.

We are all familiar with energy as a basic drive. If we are tired, we lack it; if we are leaping around with joy, we possess it. But what is energy? The energy that fires up our bodies comes from combusting chemicals, changing molecules from one type into another with energy being released in the process. But what types of energy cause a skier to speed down a slope or a light bulb to shine? Are they really the same thing?

Coming in so many different guises, energy is difficult to define. Even now, physicists do not know intrinsically what it is, even though they are expert at describing what it does and how to handle it. Energy is a property of matter and space, a sort of fuel or encapsulated drive with the potential to create, to move or to change. Philosophers of nature going back to the Greeks had a vague notion of energy as a force or essence that gives life to objects, and this idea has stuck with us through the ages.

Energy exchange It was Galileo who first spotted that energy might be transformed from one type to another. Watching a pendulum swinging back and forth, he saw that the bob exchanges height for forward motion,

timeline

c.600BC
Thales of Miletus recognizes that materials change form

AD1638
Galileo notices the exchange of kinetic and potential energy in a pendulum

1676
Leibniz formulates energy exchange mathematically and names it *vis viva*

and vice versa as the speed then brings the pendulum back up again before it falls and repeats the cycle. The pendulum bob has no sideways velocity when it is at either peak of its swing, and moves most quickly as it passes through the lowest point.

Galileo reasoned that there are two forms of energy being swapped by the swinging bob. One is gravitational potential energy, which may raise a body above the Earth in opposition to gravity. Gravitational energy needs to be added to lift a mass higher, and is released when it falls. If you have ever cycled up a steep hill you will know it takes a lot of energy to combat gravity. The other type of energy in the bob is kinetic energy – the energy of motion that accompanies speed. So the pendulum converts gravitational potential energy into kinetic energy and vice versa. A canny cyclist uses exactly the same mechanism. Riding down a steep hill, she could pick up speed and race to the bottom even without pedalling, and may use that speed to climb some of the way up the next hill (see box).

Likewise, the simple conversion of potential into kinetic energy can be harnessed to power our homes. Hydroelectric schemes and tidal barrages release water from a height, using its speed to drive turbines and generate electricity.

Many faces of energy Energy manifests as many different types that can be held temporarily in different ways. A compressed spring can store within it elastic energy that can be released on demand. Heat energy increases the vibrations of atoms and molecules in the hot material. So a metal pan on a cooker heats up because the atoms within

Energy formulae

Gravitational potential energy (PE) is written algebraically as $PE = mgh$, or mass (m) times gravitational acceleration (g) times height (h). This is equivalent to force ($F = ma$ from Newton's second law) times distance. So a force is imparting energy.

Kinetic energy (KE) is given by $KF = \frac{1}{2} mv^2$ so the amount of energy scales with the square of velocity (v). This also comes from working out the average force times the distance moved.

1807
Young names 'energy'

1905
Einstein shows that mass and energy are equivalent

it are being made to wobble faster by the input of energy. Energy can also be transmitted as electric and magnetic waves, such as light or radio waves, and stored chemical energy may be released by chemical reactions, as happens in our own digestive systems.

Einstein revealed that mass itself has an associated energy that can be released if the matter is destroyed. So, mass and energy are equivalent. This is his famous $E = mc^2$ equation – the energy (E) released by the destruction of a mass (m) is m times the speed of light (c) squared. This energy is released in a nuclear explosion or in the fusion reactions that power our Sun (see pages 136–43). Because it is scaled by the speed of light squared, which is very large (light travels at 300 million metres per second in a vacuum), the amount of energy released by destroying even a few atoms is enormous.

We consume energy in our homes and use it to power industry. We talk about energy being generated, but in reality it is being transformed from one type to another. We take chemical energy from coal or natural gas and convert it into heat that spins turbines and creates electricity. Ultimately even the chemical energy in coal and gas comes from the Sun, so solar energy is the root of everything that operates on Earth. Even though we worry that energy supplies on Earth are limited, the amount of energy that can be derived from the Sun is more than enough to power our needs, if we can only harness it.

Energy conservation Energy conservation as a rule of physics is much more than reducing our use of household energy; it states that the total amount of energy is unchanged even though it may switch between different types. The concept appeared relatively recently only after many types of energies were studied individually. At the start of the 19th century, Thomas Young introduced the word energy; before then this life force was called *vis viva* by Gottfried Leibniz who originally worked out the mathematics of the pendulum.

It was quickly noticed that kinetic energy alone was not conserved. Balls or flywheels slowed down and did not move forever. But fast motions did often cause machines to heat up by friction, such as when boring metal cannon tubes, so experimenters deduced that heat was one destination for released energy. Gradually, on accounting for all the different types of energy in built machines, the scientists began to show that energy is transferred from one type to another and is not destroyed or created.

Momentum The idea of conservation in physics is not limited to energy. Two other concepts are closely related – the conservation of linear momentum and the conservation of angular momentum. Linear momentum is defined as the product of mass and velocity, and describes the difficulty of slowing a moving body. A heavy object moving quickly has high momentum and is difficult to deflect or stop. So a truck moving at 60 kilometres an hour has more momentum than a car moving at the same speed, and would do even more damage if it hit you. Momentum has not just a size but, because of the velocity, it also acts in a specific direction. Objects that collide exchange momentum such that overall it is conserved, both in amount and direction. If you have ever played billiards or pool you have used this law. As two balls collide, they transfer motion from one to the other so as to conserve momentum. So if you hit a still ball with a moving one, the final paths of both balls will be a combination of the velocity and direction of the initial moving ball. The speed and direction of both can be worked out assuming that momentum is conserved in all directions.

Angular momentum conservation is similar. Angular momentum, for an object spinning about a point, is defined as the product of the object's linear momentum and the distance it is away from the rotation point. Conservation of angular momentum is used to effect in performances by spinning ice skaters. When their arms and legs are stretched out they whirl slowly, but just by pulling their limbs in to their body they can spin faster. This is because the smaller dimensions require an increased rotation speed to compensate. Try doing this in an office chair; it works too.

Conservation of energy and momentum are still basic tenets of modern physics. They are concepts that have found a home even in contemporary fields such as general relativity and quantum mechanics.

the condensed idea
Indestructible energy

06 Simple harmonic motion

Many vibrations adopt simple harmonic motion, which mimics the swing of a pendulum. Related to circular motion, it is seen in vibrating atoms, electrical circuits, water waves, light waves and even wobbling bridges. Although simple harmonic motion is predictable and stable, adding even small extra forces can destabilize it and may precipitate catastrophe.

Vibrations are extremely common. We have all sat down quickly and bounced for a few seconds on a well-sprung bed or chair, perhaps plucked a guitar string, fumbled for a swinging light cord or heard loud feedback from an electronic speaker. These are all forms of vibration.

Simple harmonic motion describes how an object that is pushed out of place feels a correcting force that tries to restore it. Overshooting the starting point, it wobbles back and forth until it settles back into its original position. To cause simple harmonic motion, the correcting force always opposes the object's motion and scales with the distance by which it is moved. So as the object pulls further away, it feels a stronger force pushing it back. Now moving, it is flung out the other way and, like a child on a swing, again feels a push backward that eventually stops it and sends it back. So it oscillates back and forth.

Pendulums Another way of imagining simple harmonic motion is to see it as circular motion being projected onto a line, such as the shadow of the seat of a child's swing as it appears on the ground. Like the pendulum

timeline

AD 1640
Galileo designs the pendulum clock

1851
Foucault's pendulum shows the rotation of the Earth

bob, the shadow seat, moving backwards and forwards as the seat swings, moves slowly near the ends and fast in the middle of its cycle. In both cases, the bob or seat swaps gravitational potential energy, or height, for kinetic energy, or speed.

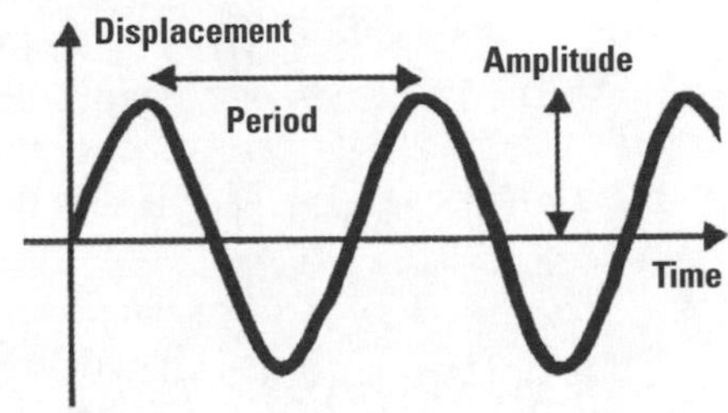

A swinging pendulum bob follows simple harmonic motion. Over time its distance from the central starting point traces out a sine wave, or a harmonic tone at the frequency of the pendulum. The bob would like to hang down vertically at rest, but once pushed to one side, the force of gravity pulls it back to the centre and adds some speed, making the oscillations persist.

Rotating Earth Pendulums are sensitive to the Earth's rotation. The spin of the Earth causes the plane of its swing to slowly turn. If you imagine a pendulum hanging above the North Pole, it swings in a plane that is fixed relative to the stars. The Earth rotates beneath it, so watching from a spot on the Earth its swinging motion seems to rotate 360 degrees in a day. There is no such rotation effect if the pendulum is hung above the equator because the pendulum rotates with the Earth, so its swing plane does not change. From any other latitude the effect lies somewhere in between. So, the fact that the Earth is rotating can be proved by simply watching a pendulum.

French physicist Léon Foucault famously devised a very public demonstration of this by hanging a huge 70-metre-high pendulum from the ceiling of the Panthéon in Paris. Today, many museums around the world also host giant Foucault pendulums. To work, their first swing needs to be set off very smoothly so that the swing plane is steady and no twists are introduced. The traditional way to do this is to tie the bob back with a string and then burn through the string with a candle to release it gently. To keep the giant pendulums moving for a long time they are often motor assisted to offset slowing due to air resistance.

1940
The Tacoma Narrows Bridge collapses

2000
The Millennium (or 'wobbly') Bridge in London is affected by resonance and closes

Time keeping Although known since the 10th century, the pendulum was not widely used in clocks until the 17th century. The time it takes a pendulum to swing depends on the length of the string. The shorter the string the faster it swings. To keep its timekeeping accurate, the length of the pendulum clock of Big Ben in London is adjusted by adding old penny coins to the heavy pendulum block. The coins change the bob's centre of mass, which is easier and more accurate to alter than moving the entire pendulum up and down.

> **'If an old English penny is added to the [Big Ben] pendulum it gains two fifths of a second a day. We have yet to work out what a Euro will do.'**
>
> **Thwaites & Reed, 2001**
> (Big Ben maintenance company)

Simple harmonic motion is not limited to pendulums but is very common throughout nature. It is seen wherever there are free vibrations, from oscillating currents in electrical circuits to the motion of particles in water waves and even the movement of atoms in the early universe.

Resonances More complicated vibrations can be described by taking simple harmonic motion as a starting point and adding extra forces. Vibrations may be boosted, by adding extra energy from a motor, or damped, by absorbing some of their energy so they diminish. For example,

Good vibrations

Electronic circuits can oscillate when currents within them flow back and forth, just like a pendulum's motion. Such circuits can make electronic sounds. One of the earliest electronic instruments is the 'theremin'. It produces eerie soaring and swooping tones and was used by the Beach Boys in their song 'Good vibrations'. The theremin consists of two electronic antennae and is played without even touching the instrument but by waving your hands near it. The player controls the pitch of the tone with one hand and the volume with the other, each hand acting as part of an electronic circuit. The theremin is named after its inventor, Russian physicist Léon Theremin, who was developing motion sensors for the Russian government in 1919. He demonstrated it to Lenin, who was impressed, and introduced it to the United States in the 1920s. Theremins were built commercially by Robert Moog who then went on to develop the electronic synthesiser that revolutionized pop music.

a cello string can be made to vibrate for a long time by regularly bowing it. Or a ringing piano string can be damped by applying a felt block to soak up its energy. Driving forces, such as bowing, may be timed to reinforce the main oscillations, or they may be out of time. If they are not synchronized, the oscillating system may start to behave surprisingly oddly very quickly.

This dramatic turnaround in behaviour sealed the fate of one of the longest bridges in the United States, Tacoma Narrows Bridge in Washington. The suspension bridge across the Tacoma Narrows acts like a thick guitar string – it vibrates easily at specific frequencies that correspond to its length and dimensions. Like a musical string, it resonates with this fundamental note but also reverberates with harmonics (multiples) of that base note. Engineers try to design bridges so that their fundamental notes are very different from the frequencies of natural phenomena, such as vibrations due to wind, moving cars or water. However, on this fateful day the engineers had not prepared enough.

Tacoma Narrows Bridge (locally known as Galloping Gertie) is a mile long and made of heavy steel girders and concrete. Nevertheless, one November day in 1940 the wind became so strong that it started to set up twisting oscillations at the bridge's resonant frequency, causing it to shake wildly and eventually rip apart and collapse. Luckily there were no fatalities, apart from one terrified dog that bit the person who tried to rescue him from a car before it tumbled off. Engineers have since fixed the bridge to stop it twisting, but even today bridges can sometimes resonate due to unforeseen forces.

Vibrations that are amplified by extra energy can get out of hand very quickly and may behave erratically. They can even become chaotic, so that they no longer follow a regular or predictable beat. Simple harmonic motion is the underlying stable behaviour, but stability is easily upset.

the condensed idea
The science of swing

07 Hooke's law

Derived originally from the stretching of springs in watches, Hooke's law shows how materials deform when forces are applied. Elastic materials stretch in proportion to the force. A prolific contributor to architecture as well as science, it is strange that Robert Hooke is remembered only for this one law. But, like its discoverer, Hooke's law crosses many disciplines, being used in engineering and construction as well as materials science.

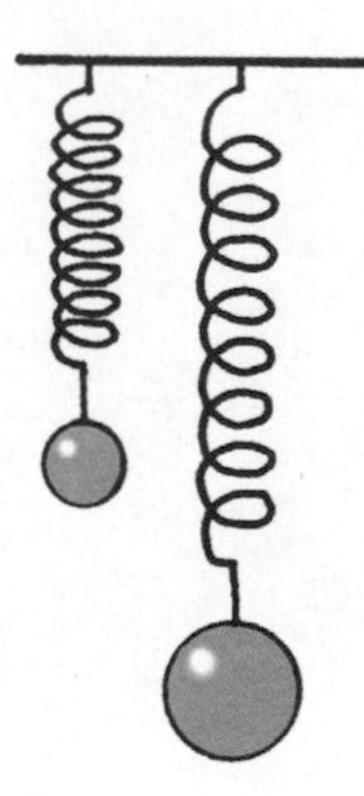

When you tell the time on your rotary watch you are indebted to Robert Hooke, the 17th-century British polymath who invented not only the balance spring and escapement mechanisms in watches and clocks but also built Bedlam and named the cell in biology. Hooke was an experimenter rather than a mathematician. He organized scientific demonstrations at the Royal Society in London and invented many devices. While working on springs he discovered Hooke's law, which says that the amount by which a spring extends is proportional to the force with which you pull it. So if you pull twice as hard then it stretches twice as far.

Elasticity Materials that obey Hooke's law are called 'elastic'. As well as stretching, elastic materials return to their original shape when any force is removed – their stretching is reversible. Rubber bands and stiff wire springs behave like this. Chewing gum on the other hand does not – it stretches when you pull it but remains stretched when you stop. Many materials behave elastically within some modest range of forces. But if they are pulled too far they may break, or fail. Other materials are too rigid or pliable to be called elastic, such as ceramics or clay.

timeline

AD 1660
Hooke discovers his law of elasticity

1773
Harrison receives an award for successful longitude measurement

ROBERT HOOKE 1635–1703

Robert Hooke was born on the Isle of Wight in England, the son of a curate. He studied at Christ Church, Oxford, working as assistant to physicist and chemist Robert Boyle. In 1660 he discovered Hooke's law of elasticity and soon after was appointed Curator of Experiments for meetings at the Royal Society. Publishing *Micrographia* five years later, Hooke coined the term 'cell', after comparing the appearance of plant cells under a microscope to the cells of monks. In 1666 Hooke helped rebuild London after the Great Fire, working with Christopher Wren on the Royal Greenwich Observatory, the Monument and Bethlem Royal Hospital (known as 'Bedlam'). He died in London in 1703 and was buried at Bishopsgate in London but his remains were moved to North London in the 19th century and their current whereabouts are unknown. In February 2006 a long lost copy of Hooke's notes from Royal Society meetings was discovered and is now housed at the Royal Society in London.

According to Hooke's law, an elastic material always requires the same amount of force to stretch it by some length. This characteristic force depends on the material's stiffness (known as its elastic modulus). A stiff material needs a large force to extend it. Materials with very high stiffness include hard substances such as diamond, silicon carbide and tungsten. More pliable materials include aluminium alloy and wood.

A material that has been stretched is said to be under a strain. Strain is defined as the percentage increase in length due to stretching. The force that is applied (per unit area) is also known as the stress. Stiffness is defined as the ratio of stress to strain. Many materials, including steel, carbon fibre and even glass, have a constant elastic modulus (for small strains), and so follow Hooke's law. When constructing buildings, architects and engineers take these properties into account so that when heavy loads are applied the structure does not stretch or buckle.

1979

The first bungee jump takes place in Bristol, UK

Bounce back Hooke's law is not just for engineers. Thousands of backpackers rely on Hooke's law every year when they try bungee jumping, leaping off a tall platform attached to an elastic cord. Hooke's law tells the jumper how much the cord will stretch when it feels the force of their weight. It is critical to do this calculation right and use the correct length of rope so that the body plunging head first towards the canyon floor bounces back before smashing into it. Bungee jumping as a sport was taken up by British daredevils who dived off Clifton Suspension Bridge in Bristol in 1979, apprently inspired by watching television pictures of Vanuatu locals jumping from great heights with tree vines tied to their ankles as a test of bravery. The jumpers were arrested, but continued leaping off bridges and spread their idea worldwide until it became a commercialized experience.

Longitude Travellers also rely on Hooke's law in another way, to help them navigate. Although measuring latitude, from north to south, is easy by monitoring the height of the Sun or stars in the sky, it is much harder to work out your longitude, or east–west location, around the Earth. In the 17th and early 18th centuries sailors' lives were in peril because of their inability to pinpoint where they were. The British government offered a cash prize of £20,000, a huge sum at the time, to someone who could overcome the technical problems of longitude measurement.

Because of the time differences as you travel from east to west across the globe, longitude can be measured by comparing your local time at sea, say at midday, with the time at some other known place, such as Greenwich in London. Greenwich lies at zero degrees longitude because time was noted relative to the observatory there; we now call it Greenwich Mean Time. This was all well and good, but how could you know the time in Greenwich if you were in the middle of the Atlantic? Just as nowadays if you flew from London to New York, you could bring a watch set to London time with you. But at the start of the 18th century, this was not easy. Clock technology at the time was not so advanced and the most accurate timepieces incorporated pendulums that were useless on a rolling ship. John Harrison, a British watchmaker, invented new devices that used rocking weights on springs instead of a dangling pendulum. But in sea tests even these failed to impress. One problem with using springs for timing was that their stretchiness changes with temperature. For ships sailing from the tropics to the poles this made them impractical.

> **'If I have seen further, it is by standing on the shoulders of giants'**
>
> **Isaac Newton**, 1675
> in a (possibly sarcastic) letter to Hooke

Harrison came up with a novel solution. He incorporated into the clock a bimetallic strip, made from two different metals bonded together. The two metals, such as brass and steel, expand by different amounts as they warm up, causing the strip to bend. Incorporated into the clock mechanism the strip compensated for the temperature changes. Harrison's new clock, called a chronometer, won the cash prize and solved the longitude problem.

All four of Harrison's experimental clocks sit today at Greenwich Observatory in London. The first three are quite large, made of brass and demonstrating intricate spring balance mechanisms. They are beautifully made and a pleasure to watch. The fourth, the winning design, is much more compact and looks just like a large pocket watch. It is less aesthetically pleasing but more accurate. Similar clocks were used for many years at sea until the arrival of the quartz electronic clock.

Hooke Hooke achieved so much, he has been called the Leonardo da Vinci of London. A key player in the scientific revolution, he contributed to many areas of science, from astronomy to biology, and even architecture. Clashing famously with Isaac Newton, the two scientists developed considerable animosity for one another. Newton was upset when Hooke refused to accept his theory of the colour of light and never credited Hooke for suggesting the inverse square theory of gravity.

It seems surprising that despite these achievements Hooke is not better known. No portraits of him survive and Hooke's law itself is a modest record for such an innovative man.

the condensed idea
Elastic fantastic

08 Ideal gas law

The pressure, volume and temperature of a gas are all linked, and the ideal gas law tells us how. If you heat a gas, it wants to expand; if you compress it, it takes up less space but has higher pressure. The ideal gas law is familiar to air travellers who shiver at the thought of the extremely cold air outside their plane, or mountaineers who expect a drop in temperature and pressure as they climb a mountain. Charles Darwin may even have blamed the ideal gas law for not cooking his potatoes when he was camping at altitude in the Andes.

If you have ever used a pressure cooker then you have used the ideal gas law to prepare your food. How do pressure cookers work? They are sealed pans that prevent the loss of steam during cooking. Because no steam escapes, as any liquid water boils the extra steam builds up and raises the pressure inside. The pressure can become high enough that it prevents further water vapour from bubbling off and allowing the temperature of the soup inside to rise above the normal boiling point of water, 100 degrees Celsius. This cooks the food faster, so it doesn't lose its flavour.

The ideal gas law is written as: $PV = nRT$ where P is pressure, V is volume, T is temperature, and n is the number of moles of gas (where 1 mole has 6×10^{23}, or Avogadro's number, of atoms in it) and R is a number called the gas constant.

The ideal gas law, first stated by French physicist Emil Clapeyron in the 19th century, tells us how the pressure, temperature and volume of a gas are all interrelated. Pressure increases if the volume is squeezed or temperature is raised. Imagine a box with air inside. If you reduced the volume of that box by half, then the pressure would be doubled. If you heated the original box to twice its temperature, then its pressure would also double.

c.350BC Aristotle states 'nature abhors a vacuum'

AD1650 Otto von Guericke builds the first vacuum pump

In deriving the ideal gas law, Clapeyron combined two earlier laws, one by Robert Boyle and another by Jacques Charles and Joseph Louis Gay-Lussac. Boyle had spotted links between pressure and volume, and Charles and Gay-Lussac between volume and temperature. Clapeyron united the three quantities by thinking about a quantity of gas called a 'mole', a term describing a certain number of atoms or molecules, namely 6×10^{23} (that's a 6 followed by 23 zeros), also known as Avogadro's number. Although this sounds like a lot of atoms, it is roughly the number of atoms that you would find in the graphite of a pencil. The mole is defined as the number of carbon-12 atoms in 12 grams of carbon. Alternatively, if you had Avogadro's number of grapefruit, they would take up the entire volume of the Earth.

> **'There is hopeful symbolism in the fact that flags do not wave in a vacuum.'**
>
> **Arthur C. Clarke, b.1917**

Ideal gas So what is an ideal gas? Simply put, an ideal gas is one that obeys the ideal gas law. It does this because the atoms or molecules that make it up are very small compared with the distances between them, so when they bounce around they scatter off one another cleanly. Also, there are no extra forces between particles that could cause them to stick together, such as electrical charges.

'Noble' gases such as neon, argon and xenon behave as ideal gases made up of individual atoms (rather than molecules). Symmetric light molecules like those of hydrogen, nitrogen or oxygen behave almost like ideal gases, whereas heavier gas molecules such as butane are less likely to.

Gases have very low densities and the atoms or molecules in them are not held together at all but are free to move around. In ideal gases, the atoms behave just like thousands of rubber balls let loose in a squash court, bouncing off each other and the walls of the container. Gases have no boundary but can be held within a container that defines a certain volume. Reducing the size of that container pushes the molecules closer together and, according to the gas law, increases both the pressure and temperature.

1662	1672	1802	1834
Boyle's law is established (PV = constant)	The Papin digester is invented	Charles's and GayLussac's law is established (V/T =constant)	Clapeyron devises the ideal gas law

The pressure of an ideal gas comes about from the forces of the atoms and molecules hitting the walls of the container, and each other, as they jostle around. According to Newton's third law (see page 10) the rebounding particles exert an opposite force on the walls. The collisions with the walls are elastic, so they bounce off without losing energy or sticking, but they transfer momentum to the box, felt as a pressure. The momentum would make the box move outwards, but its strength resists any movement, and the forces are felt in many directions, balancing out on average.

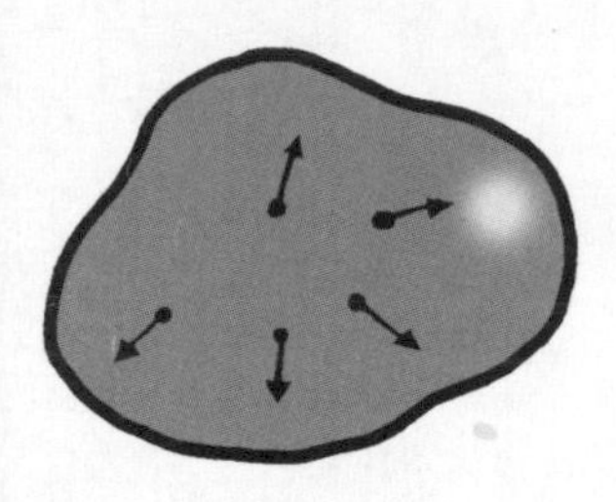

Low pressure

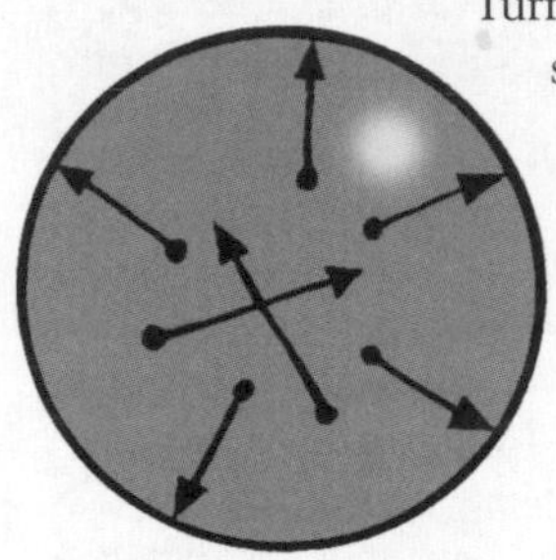

High pressure

Turning up the temperature increases the particle speeds, so the forces on the walls become even larger. Heat energy is transferred to the molecules, increasing their kinetic energy and making them move around quicker. When they hit with the walls they transfer even more momentum, again pumping up the pressure.

Reducing the volume increases the density of the gas so there are more collisions with the walls and pressure rises again. The temperature also increases because, as energy is conserved, the molecules speed up when they are in a restricted space.

Some real gases do not follow this law exactly. Gases with large or complex molecules may experience extra forces between them that mean they tend to clump together more often than in an ideal gas. Such sticky forces may arise due to the electric charges on atoms that make up the molecules and are more likely if the gas is highly compressed or very cold so the molecules are moving slowly. Really sticky molecules like proteins or fats never even become gases.

Pressure and altitude When you climb a mountain on Earth the pressure of the atmosphere drops, compared with the pressure if you were at sea level, just because there is less atmosphere above you. You may have noticed that this coincides with a drop in temperature. Flying in an aeroplane, the outside temperature drops to well below freezing. This is a demonstration of the ideal gas law.

At high altitude, because the atmospheric pressure is low, water boils at a much lower temperature than at sea level. Since food does not cook well, mountaineers sometimes use pressure cookers. Even Charles Darwin bemoaned not having one handy during his travels in the Andes in 1835, although he knew of the 'steam digester' that had been invented by French physicist Denis Papin in the late 17th century.

As Darwin wrote in his *Voyage of the Beagle*:

'At the place where we slept water necessarily boiled, from the diminished pressure of the atmosphere, at a lower temperature than it does in a less lofty country; the case being the converse of that of a Papin's digester. Hence the potatoes, after remaining for some hours in the boiling water, were nearly as hard as ever. The pot was left on the fire all night, and next morning it was boiled again, but yet the potatoes were not cooked. I found out this, by overhearing my two companions discussing the cause, they had come to the simple conclusion, "that the cursed pot [which was a new one] did not choose to boil potatoes."'

Vacuum If you could fly above the mountains to the top of the atmosphere, perhaps out into space, the pressure would drop to almost zero. A perfect vacuum would not contain any atoms, but nowhere in the universe is this true. Even in outer space there are sparsely spread atoms, numbering just a few hydrogen atoms per cubic centimetre. The Greek philosophers Plato and Aristotle did not believe that a pure vacuum could exist, as 'nothing' could not exist.

Today, the ideas of quantum mechanics have also swept aside the idea of the vacuum as empty space by suggesting it is seething with virtual subatomic particles popping in and out of existence. Cosmology even suggests space can hold a negative pressure that manifests itself as dark energy, accelerating the expansion of the universe. It seems nature truly abhors a vacuum.

the condensed idea
Pressure cooker physics

09 Second law of thermodynamics

The second law of thermodynamics is a pillar of modern physics. It says that heat travels from hot to cold bodies, and not the other way around. Because heat measures disorder, or entropy, another way of expressing the concept is that entropy always increases for an isolated system. The second law is tied to the progression of time, the unfolding of events and the ultimate fate of the universe.

When you add hot coffee to a glass of ice, the ice heats up and melts and the coffee is cooled. Have you ever asked why the temperature doesn't become more extreme? The coffee could extract heat from the ice, making itself hotter and the ice even cooler. Our experience tells us this doesn't happen, but why is this so?

The tendency of hot and cold bodies to exchange heat and move towards an even temperature is captured in the second law of thermodynamics. It says that, overall, heat cannot flow from a cold to a hot object.

So how do refrigerators work? How can we chill a glass of orange juice if we cannot transfer its warmth to something else? The second law allows us to do this in special circumstances only. As a by-product of cooling things down, refrigerators also generate a lot of heat, as you can tell if you put your hand behind the back of one. Because they liberate heat, they do not in fact violate the second law if you look at the total energy of the refrigerator and its surroundings.

AD 1150
Bhaskara proposes a perpetual motion wheel

1824
Sadi Carnot lays the foundations of thermodynamics

Entropy Heat is really a measure of disorder and, in physics, disorder is often quantified as 'entropy', which measures the ways in which a number of items can arrange themselves. A packet of uncooked spaghetti, a bundle of aligned pasta sticks, has low entropy because it shows high order. When the spaghetti is thrown into a pan of boiling water and becomes tangled, it is more disordered and so has higher entropy. Similarly, neat rows of toy soldiers have low entropy, but their distribution has higher entropy if they are scattered across the floor.

> **'Just as the constant increase of entropy is the basic law of the universe, so it is the basic law of life to be ever more highly structured and to struggle against entropy.'**
>
> **Václav Havel, 1977**

What has this got to do with refrigerators? Another way of stating the second law of thermodynamics is that, for a bounded system, entropy increases; it never decreases. Temperature is directly related to entropy and cold bodies have low entropy. Their atoms are less disordered than those in hot bodies, which jiggle around more. So any changes in the entropy of a system, considering all its parts, must produce a net effect that is an increase.

In the case of the refrigerator, cooling the orange juice decreases its entropy, but this is compensated for by the hot air that the appliance produces. In fact the entropy increase of the hot air actually exceeds any drop due to chilling. If you consider the whole system, refrigerator and surroundings, then the second law of thermodynamics still holds true. Another way of stating the second law is that entropy always increases.

The second law is true for an isolated system, a sealed one where there is no influx into or outflow of energy from it. Energy is conserved within it. The universe itself is an isolated system, in that nothing exists outside it, by definition. So for the universe as a whole, energy is conserved and entropy must always increase. Small regions might experience a slight decrease in entropy, such as by cooling, but this has to be compensated for, just like the refrigerator, by other regions heating up and creating more entropy so that the sum increases overall.

1850 Rudolf Clausius defines entropy and the 2nd law

1860 Maxwell postulates the existence of his demon

2007 Leigh claims to have built a demon machine

The (un)fashionable universe?

Astronomers recently tried to calculate the average colour of the universe, by adding up all the starlight in it, and found it is not sunshine yellow or pink or pale blue, but a rather depressing beige. In billions of years, when entropy finally wins out over gravity, the universe will become a uniform sea of beige

What does an increase in entropy look like? If you pour chocolate syrup into a glass of milk, it starts off with low entropy; the milk and syrup are distinct swathes of white and brown. If you increase the disorder by stirring the drink, then the molecules become mixed up together. The end point of maximum disorder is when the syrup is completely mixed into the milk and it turns a pale caramel colour.

Thinking again of the whole universe, the second law likewise implies that atoms gradually become more disordered over time. Any clumps of matter will slowly disperse until the universe is awash with their atoms. So the eventual fate of the universe, which starts out as a multicolour tapestry of stars and galaxies, is a grey sea of mixed atoms. When the universe has expanded so much that galaxies are torn apart and its matter is diluted, all that will remain is a blended soup of particles. This end state, presuming the universe continues to expand, is known as 'heat death'.

Perpetual motion Because heat is a form of energy, it can be put to work. A steam engine converts heat into mechanical movement of a piston or turbine, which may produce electricity. Much of the science of thermodynamics was developed in the 19th century from the practical engineering of steam engines, rather than first being deduced by physicists on paper. Another implication of the second law is that steam engines, and other engines that run off heat energy, are not perfect. In any process that changes heat into another form of energy a little energy is lost, so that the entropy of the system as a whole increases.

The idea of a perpetual motion machine, an engine that never loses energy and so can run forever, has been tantalizing scientists since medieval times. The second law of thermodynamics scotched their hopes, but before this was known many of them put forward sketches of possible machines. Robert Boyle imagined a cup that drained and refilled itself, and the Indian mathematician Bhaskara proposed a wheel that propelled its own rotation by dropping weights along spokes as it rolled. In fact, on closer inspection, both machines lose energy. Ideas like these were so widespread that even in the 18th century perpetual motion machines garnered a bad name. Both the French Royal Academy of Sciences and the American Patent Office banned consideration of perpetual motion machines. Today they remain the realm of eccentric backyard inventors.

Another view of the laws of thermodynamics

First law
You can't win
(see Conservation of energy, page 20)

Second law
You can only lose
(see page 36)

Third law
You can't get out of the game
(see Absolute zero, page 40)

Maxwell's demon One of the most controversial attempts to violate the second law was proposed as a thought experiment by the Scottish physicist James Clerk Maxwell, in the 1860s. Imagine two boxes of gas, side by side, both at the same temperature. A small hole is placed between the boxes, so that particles of gas can pass from one box to the other. If one side was warmer than the other, particles would pass through and gradually even out the temperature. Maxwell imagined that there was a tiny demon, a microscopic devil, who could grab only fast molecules from one box and push them through into the other. In this way the average speed of molecules in that box would increase, at the expense of the other. So, Maxwell postulated, heat could be moved from the colder to hotter box. Wouldn't this process violate the second law of thermodynamics? Could heat be transferred into the hotter body by selecting the right molecules?

An explanation of why Maxwell's demon could not work has puzzled physicists ever since. Many have argued that the process of measuring the particles' velocities and opening and closing any trap door would require work and therefore energy, so this would mean that the total entropy of the system would not decrease. The nearest anyone has come to a 'demon machine' is the nanoscale work of Edinburgh physicist David Leigh. His creation has indeed separated fast- and slow-moving particles, but requires an external power source to do so. Because there is no mechanism that could move particles without using extra energy, even today's physicists have not found a way to violate the second law. Thus far, at least, it is holding fast.

the condensed idea
Law of disorder

10 Absolute zero

Absolute zero is the imagined point at which a substance is so cold its atoms cease moving. Absolute zero itself has never been reached, neither in nature nor in the laboratory. But scientists have come very close. It may be impossible to get to absolute zero, and even if we did we might not know because no thermometer could measure it.

When we measure the temperature of something we are recording the average energy of the particles that make it up. Temperature indicates how quickly the particles are vibrating or moving around. In a gas or liquid, molecules are free to travel in any direction, and often bounce off one another. So temperature is related to the average speed of the particles. In a solid, atoms are anchored in a lattice structure, like Meccano held together by electronic bonds. When this becomes hot, the atoms are energetic and jiggle around a lot, like wobbly Jello, while sitting in their positions.

As you cool a material, its atoms move around less. In a gas their speeds drop; in a solid their vibrations are reduced. As the temperature drops further and further, atoms move less and less. If cooled enough, a substance could become so cold that its atoms stop moving completely. This hypothetical still point is called absolute zero.

Kelvin scale The idea of absolute zero was recognized in the 18th century by extrapolating a graph of temperature and energy to zero. Energy rises steadily with temperature, and the line connecting the two quantities can be projected backwards to find the temperature at which the energy reaches zero: −273.15 degrees Celsius or −459.67 degrees Fahrenheit.

timeline

AD 1702 Guillaume Amontons proposes the idea of absolute zero

1777 Lambert proposes an absolute temperature scale

1802 Gay-Lussac identifies absolute zero at -273 Celsius

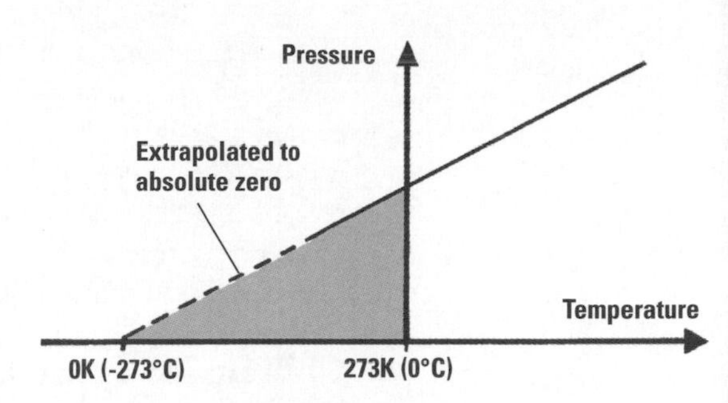

In the 19th century, Lord Kelvin proposed a new temperature scale that started at absolute zero. Kelvin's scale effectively took the Celsius temperature scale and shifted it. So, instead of water freezing at 0 degrees Celsius it does so at 273 kelvins and boils at 373 kelvins (equivalent to 100 degrees Celsius). The upper ranges of this scale are fixed, as is the triple point of water, the temperature (at a particular pressure) at which water, steam and ice can all coexist, which occurs at 273.16 kelvins or 0.01 Celsius at low pressure (less than 1% of atmospheric pressure). Today, most scientists use kelvins to measure temperature.

Big freeze How cold does absolute zero feel? We know what it feels like when the temperature outside reaches freezing or when it starts to snow. Your breath freezes and your fingers start to become numb. That's chilly enough. Parts of North America and Siberia may reach 10 or 20 degrees below this in winter, and it can reach −70 degrees Celsius at the South Pole. The coldest natural temperature experienced on Earth is a chilly −89 degrees Celsius, or 184 kelvins, witnessed at Vostok in the heart of Antarctica in 1983.

> **'Since I like to keep my popsicles at absolute zero, I employ Kelvin more than most Americans. I find that desserts aren't delicious unless they exhibit no molecular movement whatsoever.'**
>
> **Chuck Klosterman, 2004**

The temperature also drops if you climb up a mountain or fly high in the atmosphere in an aeroplane. Moving on out into space, it is even colder. Even in the deepest and emptiest reaches of space the coldest atoms have temperatures a few degrees above absolute zero. The coldest environment found so far in the universe is situated inside the Boomerang Nebula, a dark gas cloud lying just one degree above absolute zero.

1848	1900	1930	1954
The Kelvin temperature scale is defined	Kelvin delivers his 'two clouds' talk	Experimental measurements pinpoint absolute zero more precisely	Absolute zero is defined officially as -273.15 degrees Celsius

Outside this nebula, and throughout empty space, the ambient temperature is a relatively balmy 2.7 kelvins. This tepid bath is due to the cosmic microwave background radiation, heat left over from the big bang itself, which pervades all of space (see page 182). To become any cooler, regions must be shielded from this background warmth and any atoms should have lost their residual heat. So it is practically inconceivable that any location in space truly lies at absolute zero.

Chill inside Even colder temperatures have been reached temporarily in the laboratory, where physicists have tried to approach absolute zero for short periods of time. They have come very close, much closer than in ambient outer space.

Many liquid gas coolants are used in laboratories, but these are warmer than absolute zero. It is possible to cool nitrogen until it becomes a liquid at 77 kelvins (−196 degrees Celsius). Liquid nitrogen is easy to transport in cylinders and is used in hospitals for preserving biological samples, including freezing embryos and sperm in fertility clinics, as well as in advanced electronics. When cooled by dipping in liquid nitrogen, a carnation flower head becomes so brittle it fractures like porcelain when dropped on the floor.

Liquid helium is even colder, just 4 kelvins, but still well above absolute zero. By mixing two types of helium, helium-3 and helium-4, it is possible to cool the mixture to a few thousandths of a kelvin.

‘During the first half of Thomson’s career he seemed incapable of being wrong while during the second half of his career he seemed incapable of being right.’

C. Watson, 1969
(Lord Kelvin’s biographer)

To reach even colder temperatures, physicists need even cleverer technology. In 1994 at the American National Institute for Standards and Technology (NIST) in Boulder, Colorado, scientists managed to cool caesium atoms using lasers to within 700 billionths of a kelvin from absolute zero. Nine years later, scientists at the Massachusetts Institute of Technology went further still, reaching within 0.5 billionths of a kelvin.

Really, absolute zero is an abstract idea. It has never been achieved in a laboratory nor has it

LORD KELVIN 1824–1907

The British physicist Lord Kelvin, born William Thomson, addressed many problems in electricity and heat, although he is most famous for helping to build the first transatlantic submarine cable for the transmission of telegraphs. Thomson published more than 600 papers and was elected President of the prestigious Royal Society of London. He was a conservative physicist and refused to accept the existence of atoms, opposed Darwin's theories of evolution and related theories of the ages of the Earth and Sun, putting him on the losing side in many arguments. Thomson was named Baron Kelvin of Largs, after the River Kelvin that runs through Glasgow University and his home town of Largs on the Scottish coast. In 1900, Lord Kelvin gave a now-famous lecture to the Royal Institution of Great Britain where he bemoaned the fact that the 'beauty and clearness of theory' was overshadowed by 'two clouds', namely the then flawed theory of black-body radiation and the failed attempt to observe an 'ether' or gas medium through which it was assumed light travelled. The two problems he singled out would later be addressed by relativity and quantum theory, but Thomson struggled to solve them with the Newtonian physics of his day.

ever been measured in nature. As they try to approach ever closer, scientists must accept that absolute zero may never in fact be reached with certainty.

Why should this be? First of all, any thermometer that was not itself at absolute zero would add heat and so ruin the attainment of it. Secondly, it is hard to measure temperature at all at such low energies, where other effects such as superconductivity and quantum mechanics intervene and affect the motions and states of the atoms. So we can never know for sure that we have got there. For absolute zero, it may be is a case of 'there's no there there'.

the condensed idea
The big chill

11 Brownian motion

Brownian motion describes the jerky movements of small particles as they are buffeted by invisible water or gas molecules. The botanist Robert Brown first spotted it as the twitching of pollen particles on his wet microscope slides, but it was Albert Einstein who described it mathematically. Brownian motion explains how pollution diffuses through still air or water and describes many random processes, from floods to the stock market. Its unpredictable steps are linked to fractals.

The 19th-century botanist Robert Brown was looking at pollen grains under a microscope when he noticed that they did not sit still but jerked around. For a moment he wondered if they were alive. Clearly they were not, but were instead being knocked around by the motions of molecules within the water that Brown had used to coat the glass slides. The pollen particles moved in random directions, sometimes a little and occasionally quite a lot, and gradually shuffled across the slide following tracks that could not be predicted. Other scientists puzzled over Brown's discovery, which was named Brownian motion after him.

Random walk Brownian motion happens because a tiny pollen particle receives a little kick every time a water molecule bumps into it. The invisible water molecules are moving around and colliding with each other all the time, and so they regularly bump into the pollen, jostling it.

timeline

c.420BC Democritus postulates the existence of atoms

AD1827 Brown observes pollen's motion and proposes the mechanism

Even though the pollen grain is hundreds of times bigger than a water molecule, because the pollen is being hit at any instant by many molecules, each moving in random directions, there is usually a force imbalance which makes it move a little. This happens again and again and so the buffeted pollen gain follows a jagged path, a bit like the route of a staggering drunk. The pollen's path cannot be predicted in advance because the water molecules collide at random and so the pollen may dart off in any direction.

Brownian motion affects any small particle suspended in a liquid or gas. It is exhibited by even quite large particles such as smoke particles that jitterbug in air if viewed through a magnifying lens. The size of the knocks that the particle receives, depends on the momentum of the molecules. So greater buffeting is seen when the molecules of the liquid or the gas are heavy, or when they are moving fast, for instance if the fluid is hot.

The mathematics behind Brownian motion was pursued in the late 19th century but it was Einstein who brought it to the attention of physicists in his paper of 1905, the same year he published his theory of relativity and an explanation of the photoelectric effect that won him his Nobel Prize. Einstein borrowed the theory of heat, which also was based on molecular collisions, to explain successfully the exact motions that Brown had observed. On seeing that Brownian motion provided evidence for the existence of molecules in fluids, physicists were compelled to accept the theory of atoms, which was still being questioned even as late as the beginning of the 20th century.

The 'random walk' of Brownian motion

Diffusion Over time, Brownian motion may cause particles to move by quite some distance, but never so far as if their paths were unimpeded and they moved in straight lines. This is because the randomness is just as likely to send a particle back on itself as move it forwards. So if a group of particles was dropped in one spot into some liquid it would diffuse

1905
Einstein determines the mathematics behind Brownian motion

1960s
Mandelbrot discovers fractals

outwards even if nobody stirred or there were no currents in the liquid. Each particle would trundle off in its own way, causing the concentrated droplet to spread out into a diffuse cloud. Such diffusion is important for the spread of pollution from a source, such as aerosols in the atmosphere. Even if there is no wind at all the chemicals will diffuse due to Brownian motion alone.

Fractals The path followed by a particle undergoing Brownian motion is an example of a fractal. Each step of the path can be of any size and in any direction, but there is some overall pattern that emerges. This pattern has structure within it on all scales, from the tiniest imaginable to quite large modulations. This is the defining characteristic of a fractal.

Fractals were defined by Benoit Mandelbrot in the 1960s and 70s as a way of quantifying self-similar shapes. Short for fractional dimensions, fractals are patterns that look essentially the same at any scale. If you zoom in on a small piece of the pattern it looks indistinguishable from the larger scale one, so you cannot tell what the magnification is just by looking at it. These repeating and scale-less patterns appear frequently in nature, such as in the crinkles of a coastline, the branches of a tree, the fronds of a fern, or the six-fold symmetry of a snowflake.

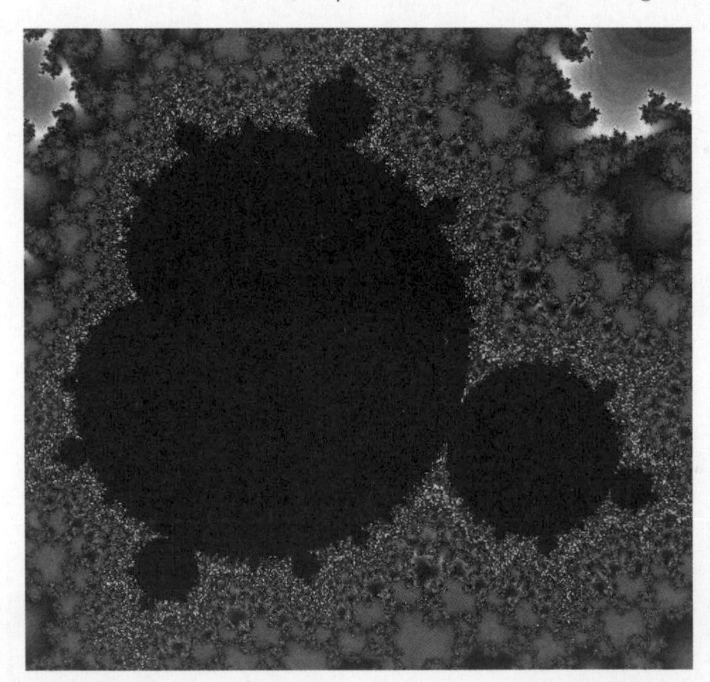

Fractional dimensions arise because their length or dimension depends on the scale at which you look. If you measure the distance between two towns along a coast you might say it is 30 kilometres between Land's End and Mount's Bay, but if you considered all the individual rocks and measured around each of them with a piece of string, you might need a piece of string a hundred kilometres long to do this. If you went further and measured around every grain of sand on the coast you might need a piece of string hundreds of kilometres long. So the absolute length here depends on the scale on which you are measuring. If you blur everything down to a coarse level then you return to your familiar 30 kilometres. In this sense, fractal dimensions measure the roughness of something, be it a cloud, a tree or a range of mountains. Many of these fractal shapes, such as the outline of a coastline, can be produced by a series of random motion steps, hence their link to Brownian motion.

The mathematics of Brownian motion, or a sequence of random movements, can be used to generate fractal patterns that are useful in many areas of science. They can create rough hewn virtual landscapes of mountains, trees and clouds for computer games or be used in spatial mapping programs that might help robots steer themselves across rough terrain, by modelling its ridges and crevices. Doctors find them helpful in medical imaging when they need to analyse the structure of complex parts of the body, such as the lungs, where branching structures run from the coarse to fine scale.

Brownian motion ideas are also useful for predicting risks and events in the future that are the summed outcome of many random events, such as floods and stock market fluctuations. The stock market can be treated as a portfolio of stocks whose prices vary randomly like the Brownian motion of a set of molecules. Brownian motion also figures in the modelling of other social processes such as those in manufacturing and decision making. The random movements of Brownian motion have had a wide influence and appear in many guises, not just in the dance of the leaves in a nice hot cup of tea.

the condensed idea
An invisible microscopic dance

12 Chaos theory

Chaos theory declares that tiny changes in circumstance can have major ramifications later on. If you left the house 30 seconds late, then even though you just missed your bus you may also have met someone who directed you to a new job, changing the path of your life for ever. Chaos theory applies most famously to the weather, where a tiny wind eddy can seed a hurricane on the other side of the planet, the so-called 'butterfly effect'. However, chaos is not chaotic in the literal sense, as patterns do emerge from it.

The fluttering of a butterfly in Brazil can cause a tornado in Texas. So says chaos theory. Chaos theory recognizes that some systems can produce very diverse behaviours even though they have very similar starting points. Weather is one such system. A very tiny tweak to temperature or pressure in one place can set off a chain of further events that can trigger a downpour somewhere else.

Chaos is somewhat of a misnomer. It is not chaotic in the sense of being utterly wild, unpredictable or unstructured. Chaotic systems are deterministic, so if you know the exact starting point they are predictable and also reproducible. Simple physics describes the series of events that unfolds, which is the same every time you try. But if you take one final outcome it is then impossible to go backwards and say from where it came, as several paths may have led to it. This is because the differences between the conditions that triggered one outcome and another may be tiny, even unmeasurable. So, divergent outcomes result from tiny input shifts. Because of this divergence, if you are uncertain about the input value then

AD 1898
Hadamard's billiards exhibit chaotic behaviour

the range of consequent behaviours may be vast. In terms of weather, if the temperature of the wind eddy is different by just a fraction of a degree from what you think it is, then your predictions may be completely wrong and you could end up not with a violent storm but rather just a light shower, or a fierce tornado in the neighbouring town. Weather forecasters are therefore limited in how far ahead they can model the climate. Even with vast amounts of data on the state of the atmosphere, from swarms of satellites orbiting the Earth and weather stations on it, forecasters can predict weather patterns for only a few days ahead. After that the uncertainties become too great due to chaos.

Development Chaos theory was developed in earnest in the 1960s by the American mathematician and meteorologist Edward Lorenz. Whilst using a computer to model the weather, Lorenz noticed that his code produced vastly different weather pattern outputs just because the input numbers were rounded up differently. To help his computations he had split up his simulations into chunks and tried to restart them in the middle by printing out numbers and then typing them in again by hand. In the printout the numbers were rounded up to three decimal places, which he typed in, but the computer's memory was handling numbers with six decimal places. So when 0.123456 was replaced with the shortened 0.123 in the middle of the simulation, Lorenz saw that the weather that resulted was completely different. Tiny errors introduced by the computer rounding had a profound effect on the final climate prediction. His models were reproducible, so not random, but the differences were difficult to interpret. Why did a tiny tweak in his code produce nice clear weather in one simulation or catastrophic storms in another?

Looking in more detail he saw that the output weather patterns were restricted to a certain set, which he called an attractor. It was not possible to produce just any type of weather by varying the inputs, but rather a set of weather patterns was favoured even if it was difficult to predict in

1961
Lorenz works on weather forecasting

2005
Neptune's moons are found to orbit chaotically

The butterfly effect

The main idea of chaos, that small changes can have big ramifications later on, is often referred to as the 'butterfly effect' after Lorenz's vision of the creature flapping its wings and causing a tornado. This idea, especially involving time travel, has been used widely in films and popular culture, including a film called *The Butterfly Effect* and even in *Jurassic Park.* In the 1946 film *It's a Wonderful Life*, the main character George is shown by an angel how his home town would have been a more miserable place had he not been born. The angel says: 'You've been given a great gift, George: a chance to see what the world would be like without you.' George finds out that his very existence saved a man from drowning, and his really is a wonderful life.

advance exactly which one should follow from the input numbers. This is a key feature of chaotic systems – they follow patterns overall but a specific end point cannot be projected back to a particular initial input because of overlapping potential paths to those outcomes. There are many different ways to get to the final output.

The connections between input and output can be traced as a graph to show the range of behaviours that a particular chaotic system can exhibit. Such a graph maps the attractor solutions, which are sometimes referred to as 'strange attractors'. A famous example is the Lorenz attractor which looks like a series of overlapping figures of eight slightly shifted and distorted from one another, mirroring the shape of butterflies' wings.

Chaos theory emerged around the same time that fractals were uncovered. In fact the two are closely related. The attractor maps of chaos solutions for many systems can appear as fractals, where the fine structure of the attractor has structure within it on many scales.

Early examples Although the availability of computers really jump started chaos theory, by allowing mathematicians to calculate behaviours repeatedly for different input numbers, simpler systems showing chaotic behaviour were recognized much earlier. For example, at the end of the 19th century, chaos was known to apply to the paths of billiard balls and the stability of orbits.

> **'Every man on that transport died! Harry wasn't there to save them because you weren't there to save Harry! You see, George: you've really had a wonderful life. Don't you see what a mistake it would be to throw it away?'**
>
> **It's a Wonderful Life, 1946**

Jacques Hadamard studied the mathematics of the motion of a particle on a curved surface, such as a ball on a golf course, known as Hadamard's billiards. On some surfaces the tracks of the particles became unstable, and they fell off the edge. Others remained on the baize but followed variable paths. Soon after, Henri Poincaré also found non-repeating solutions to the orbits of three bodies under gravity, such as the Earth with two moons, again seeing orbits that were unstable. The three bodies orbited each other in ever-changing loops but did not fly apart. Mathematicians then tried to develop this theory of multi-body motions, known as ergodic theory, and applied it to turbulent fluids and electrical oscillations in radio circuits. From the 1950s onwards, chaos theory developed rapidly as new chaotic systems were found and digital computers were introduced to ease calculations. ENIAC, one of the first computers, was used for weather forecasting and investigating chaos.

Chaotic behaviour is widespread in nature. As well as affecting the weather and other fluid flows, chaos occurs for many multiple body systems, including the orbits of planets. Neptune has more than a dozen moons. Rather than following the same orbits year after year, chaos causes Neptune's moons to ricochet around in unstable orbits that change year by year. Some scientists think that the orderly arrangement of our own solar system may be ultimately down to chaos. If our planets and others were involved in a giant game of billiards billions of years ago, shaking up all the orbits until the unstable bodies were lost altogether, then the stable pattern of planets we see today is what was left behind.

the condensed idea
Order in chaos

13 Bernoulli equation

The relationship between the speed and pressure of flowing fluids is given by Bernoulli's equation. It governs why planes fly, how blood flows through our bodies and how fuel is injected into car engines. Fast flowing fluids create low pressure that explains the lift associated with an aircraft wing and the narrowing of a jet of water coming from a tap. Using this effect to measure blood pressure, Daniel Bernoulli himself inserted tubes directly into his patients' veins.

When you run a tap, the column of water that flows from it is narrower than the aperture of the tap itself. Why is this? And how is this related to how planes fly and angioplasties?

Dutch physicist and medical doctor Daniel Bernoulli understood that moving water creates low pressure. The faster it flows, the lower its pressure. If you imagine a clear glass tube lying horizontally with water pumped through it, you can measure the pressure of that water by inserting a clear capillary tube vertically into the first pipe and watching how the height of water in the smaller tube changes. If the pressure of the water is high, the water level in the capillary rises. If it is low it drops.

When Bernoulli increased the speed of the water in the horizontal tube he observed a drop in pressure in the vertical capillary tube – this pressure drop proved proportional to the velocity of the water squared. So any flowing water, or fluid, has a lower pressure than still water. Water flowing

timeline

AD 1738

Bernoulli discovers that an increase in fluid speed causes a decrease in its pressure

from a tap has low pressure compared to the still air around it and so it is sucked into a narrower column. This applies to any fluid, from water to air.

Blood flow Trained in medicine, Bernoulli himself was fascinated by the flow of blood though the human body and invented a tool to be able to measure blood pressure. A capillary tube, inserted into a blood vessel, was used for nearly two hundred years to measure blood pressure in live patients. It must have been a relief for all concerned to find a less invasive method.

Just like water in a pipe, blood in an artery is pumped away from the heart along a pressure gradient that is set up along the length of the vessel. If an artery is narrowed, then the speed of blood flowing through the constriction increases according to Bernoulli's equation. If the vessel is half as narrow, then the blood that flows through it is four times faster (two squared). This quickening of blood flow through restricted arteries can cause problems. First, the flow may become turbulent, if its speed is fast enough, and eddies may be produced. Turbulence near the heart produces heart murmurs with a characteristic sound that doctors can recognize. Also, the pressure drop in the constricted area may suck in the soft artery wall, further aggravating the problem. If the artery is expanded, with an angioplasty, the volume of flow will increase again and all will be well.

Lift The drop in pressure with fluid speed has other profound consequences. Aeroplanes fly because the air rushing past a plane's wing also produces a pressure drop. Aircraft wings are shaped so that the top edge is more curved than the lower edge. Because of the longer path over the top, air moves faster over the top surface of the wing so the pressure there is lower than on the underside. The pressure difference gives the wing lift and allows the plane to fly. But a

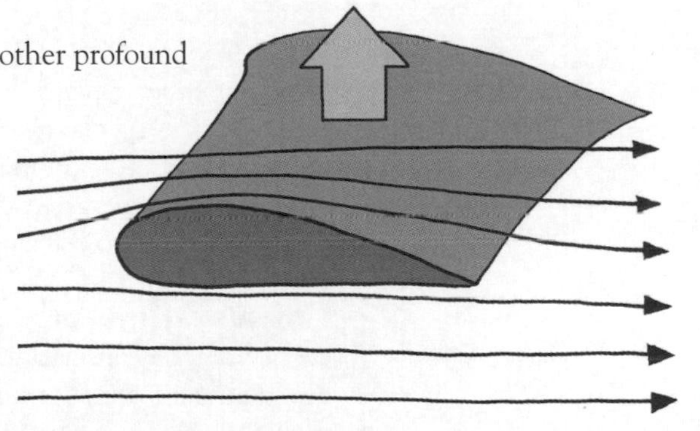

1896
A noninvasive technique is invented to measure blood pressure

1903
The Wright brothers, with aircraft wings inspired by Bernoulli, fly the first plane

DANIEL BERNOULLI 1700–82

Dutch physicist Daniel Bernoulli trained as a medic to fulfil his father's wishes, but really loved mathematics. His father Johann was a mathematician but tried to dissuade Daniel from following in his footsteps, and competed with his son throughout his career. Bernoulli completed his medical studies in Basel, but in 1724 he became a mathematics professor in St Petersburg. Working with the mathematician Leonhard Euler on fluids, he linked speed and pressure by experimenting with pipes that were eventually used by doctors to measure blood pressure by inserting them into arteries. Bernoulli realized that the flow and pressure of fluids was linked to the conservation of energy and showed that if the velocity increases then the pressure falls. Daniel won a position back in Basel in 1733, but Johann was still jealous of his son's achievements. He hated having him in the same department and even banned him from his house. Despite all this Daniel dedicated his book *Hydrodynamica*, written in 1734 but not published until 1738, to his father. But Bernoulli senior stole Daniel's ideas, publishing a similar book called *Hydraulics* soon after. Daniel, upset at the plagiarism, moved back into medicine for the rest of his career.

heavy plane has to move very fast to gain enough pressure difference to provide lift to take off.

A similar effect explains how fuel is injected into a car engine through a carburettor. A special nozzle, called a venturi tube (a wide tube with a narrower 'waist' region in its middle), produces low pressure air, by restricting and then releasing the flow, that sucks up fuel and so delivers a fuel–air mixture to the engine.

Conservation Daniel Bernoulli came to his understanding by thinking about how the conservation of energy applied to fluids. Fluids, including liquids and air, are continuous substances that can constantly deform. But they must follow the basic laws of conservation, not only of energy but also of mass and momentum. Because any moving fluid is essentially always rearranging the atoms within itself, these atoms must follow the laws of motion derived by Newton and others. So in any fluid description, atoms cannot be created or destroyed, but rather are moved around. Their collisions with one another must be considered, and when they do so their velocities are predicted by the conservation of linear momentum. Also, the total amount of energy taken up by all the particles must be fixed, and can only be moved around within the system.

> **'Heavier-than-air flying machines are impossible. I have not the smallest molecule of faith in aerial navigation other than ballooning, or of the expectation of good results from any of the trials we hear of.'**
>
> **Lord Kelvin, 1895**

These physical laws are used today to model fluid behaviour as diverse as weather patterns, ocean currents, the circulation of gas in stars and galaxies and fluid flow in our bodies. Weather prediction relies on the computer modelling of the movements of many atoms together with thermodynamics to account for changes in heat as the atoms move and change density, temperature and pressure regionally. Again, pressure changes and velocities are linked, as they cause winds to flow from high to low pressure. These same ideas were used to model the path of Hurricane Katrina as it sped towards the American coast in 2005.

The conservation laws are embodied in a series of further equations called the Navier–Stokes equations, after the scientists who devised them. They also accommodate the effects of fluid viscosity, the stickiness of the fluid, due to forces between the molecules that make it up. Dealing with conservation rather than absolute prediction, these equations track the changes and circulation of the fluid particles on average rather than following the total numbers of atoms.

The Navier–Stokes equations of fluid dynamics, although detailed enough to explain many complex systems such as climate phenomena including El Niño and hurricanes, are not able to describe very turbulent flow such as the crashing cascade of a waterfall or the flow of a fountain. Turbulence is the random motion of disturbed water, characterized by eddies and instability. It sets in when flows become very rapid and destabilize. Because turbulence is so difficult to describe mathematically, major money prizes are still being offered for scientists to come up with new equations to describe these extreme situations.

the condensed idea
Arteries and aerodynamics

14 Newton's theory of colour

We've all wondered at the beauty of a rainbow – Isaac Newton explained how they form. Passing white light through a glass prism, he found it split into rainbow hues and showed that the colours were embedded in the white light rather than imprinted by the prism. Newton's colour theory was contentious at the time but has influenced generations of artists and scientists since.

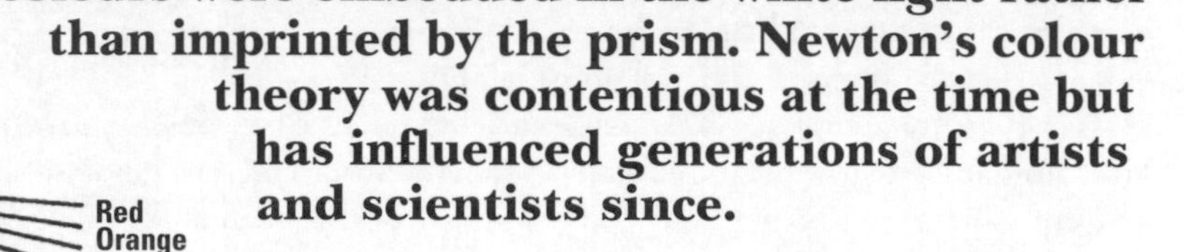

Shine a beam of white light through a prism and the emerging ray spreads out into a rainbow of colours. Rainbows in the sky appear in the same way as sunlight is split by water droplets into the familiar spectrum of hues: red, orange, yellow, green, blue, indigo and violet.

All in the mix Experimenting with light and prisms in his rooms in the 1660s, Isaac Newton demonstrated that light's many colours could be mixed together to form white light. Colours were the base units rather than being made by later mixing or by the prism glass itself, as had been thought. Newton separated beams of red and blue light and showed that these single colours were not split further if they were passed through more prisms.

Although so familiar today, Newton's theory of colour proved contentious at the time. His peers argued vociferously against it, preferring to believe

timeline

AD 1672
Newton explains the rainbow

instead that colours arose from combinations of white light and darkness, as a type of shadow. Newton's fiercest battles were with his equally famous contemporary, Robert Hooke. The pair fought publicly over colour theory throughout their lifetimes. Hooke believed instead that coloured light was an imprint, just as if you look through stained glass. He cited many examples of unusual coloured light effects in real life to back up his claim and criticized Newton for not performing more experiments.

> **'Nature and nature's laws lay hid in night;**
> **God said "Let Newton be" and all was light.'**
>
> **Alexander Pope**, 1727 (Newton's epitaph)

Newton also realized that objects in a lit room appear coloured because they scatter or reflect light of that colour, rather than colour being somehow a quality of the object. A red sofa reflects primarily red light, and a green table reflects green light. A turquoise cushion reflects blue and a little yellow light. Other colours arise from mixtures of these basic types of light.

Light waves For Newton, understanding colour was a means of interrogating the physics of light itself. Experimenting further, he concluded that light behaves in many ways like water waves. Light bends around obstacles in a similar way to sea waves around a harbour wall. Light beams can also be added together to reinforce or cancel out their brightness, as overlapping water waves do. In the same way that water waves are large-scale motions of invisible water molecules, Newton believed that light waves were ultimately ripples of miniscule light particles, or 'corpuscles', which were even smaller than atoms. What Newton did not know, until its discovery centuries later, was that light waves are in fact electromagnetic waves – waves of coupled electric and magnetic fields – and not the reverberation of solid particles. When the electromagnetic wave behaviour of light was discovered, Newton's corpuscle idea was put on ice. It was resurrected, however, in a new form when Einstein showed that light may also behave sometimes like a stream of particles that can carry energy but have no mass.

1810
Goethe publishes his treatise on colour

1905
Einstein shows light can behave as particles in some circumstances

Wave motions appear in many guises. There are two basic wave types: longitudinal and transverse waves. Longitudinal, or compression, waves result when the pulses that produce the wave act along the same direction in which the wave travels, causing a series of high and low pressure crests. Sound waves, caused for example by a drum skin vibrating in air, are longitudinal, as are the ripples of a millipede's legs as they crunch up close and then spread apart as the creature shuffles forwards. Light and water waves, on the other hand, are transverse waves where the original disturbance acts at a right angle to the direction of travel of the wave. If you sweep one end of a slinky spring from side to side a transverse wave will travel along the length of the spring even though your hand's motion is perpendicular to it. Similarly, a snake makes a transverse wave as it slithers, using side to side motion to propel it forwards. Water waves are also transverse because individual water molecules float up and down whereas the wave itself travels towards the horizon. Unlike water waves, the transverse motion of light waves is due to changes in the strength of electric and magnetic fields that are aligned perpendicular to the direction of wave propagation.

Across the spectrum The different colours of light reflect the different wavelengths of these electromagnetic waves. Wavelength is the measured distance between consecutive crests of a wave. As it passes through a prism, the white light separates into many hues because each hue is associated with a different wavelength and so they are deflected to varying degrees by the glass. The prism bends the light waves by an angle that depends on the wavelength of light, where red light is bent least and blue most, to produce the rainbow colour sequence. The spectrum of visible light appears in order of wavelength, from red with the longest through green to blue with the shortest.

colour wheel

Newton arranged the colours of the rainbow in order from red to blue and painted them onto a circular colour wheel, so he could show the ways in which colours combined. Primary colours – red, yellow and blue – were spaced around it, and when combined in different proportions could make all the other colours in between. Complementary colours, such as blue and orange, were placed opposite one another. Many artists became interested in Newton's colour theory and especially in his colour wheel that helped them depict contrasting hues and illumination effects. Complementary colours achieved maximum contrast, or were useful for painting shadows.

What lies at either end of the rainbow? Visible light is just one part of the electromagnetic spectrum. It is so important to us because our eyes have developed to use this sensitive part of the spectrum. As the wavelengths of visible light are on roughly the same scale as atoms and molecules (hundreds of billionths of a metre), the interactions between light and atoms in a material are large. Our eyes have evolved to use visible light because it is very sensitive to atomic structure. Newton was fascinated by how the eye worked; he even stuck a darning needle round the back of his own eye to see how pressure affected his perception of colour.

Beyond red light comes infrared, with wavelengths of millionths of a metre. Infrared rays carry the Sun's warmth and are also collected by night-vision goggles to 'see' the heat from bodies. Longer still are microwaves, with millimetre-to-centimetre wavelengths, and radio waves, with wavelengths of metres and longer. Microwave ovens use microwave electromagnetic rays to spin water molecules within food, heating them up. At the other end of the spectrum, beyond blue, comes ultraviolet light. This is emitted by the Sun and can damage our skin, although much of it is stopped by the Earth's ozone layer. At even shorter wavelengths are X-rays – used in hospitals because they travel through human tissue – and at the smallest wavelengths are gamma rays.

Developments While Newton elucidated the physics of light, philosophers and artists remained interested in our perception of colours. In the 19th century, German polymath Johann Wolfgang von Goethe, investigated how the human eye and mind interprets colours placed next to one another. Goethe introduced magenta to Newton's colour wheel (see box) and noticed that shadows often take on the opposite colour of the illuminated object, so that a blue shadow appears to fall behind a red object. Goethe's updated colour wheel remains the choice for artists and designers today.

the condensed idea
Beyond the rainbow

15 Huygens' principle

If you drop a stone into a pond, it produces a circular expanding ripple. Why does it expand? And how might you predict its behaviour if it then flows round an obstacle, such as a tree stump, or reflects back from the edge of the pond? Huygens' principle is a tool for working out how waves flow by imagining that every point on a wavefront is a new ripple source.

Dutch physicist Christiaan Huygens devised a practical way for predicting the progression of waves. Let's say you have cast a pebble into a lake, and rings of ripples result. If you imagine freezing a circular ripple at a moment in time, then each point on the circular wave can be thought of as a new source of circular waves whose properties match those of the frozen ripple. It is as if a ring of stones was dropped simultaneously into the water following the outline of the first wave. This next set of disturbances widens the ripple further, and the new locus marks the starting points for another set of sources of spreading wave energy. By repeating the principle many times the evolution of the wave can be tracked.

Step by step The idea that every point on a wavefront acts like a new source of wave energy with matching frequency and phase is called Huygens' principle. The frequency of a wave is the number of wave cycles that occur in some time period and the phase of a wave identifies where you are in the cycle. For example, all wave crests have the same phase, and all troughs are half a cycle away from them. If you imagine an ocean wave, the distance between two wave peaks, known as its wavelength, is maybe

timeline

AD 1655
Huygens discovers Titan

1678
Huygens' treatise on the wave theory of light is published

CHRISTIAAN HUYGENS 1629–95

Son of a Dutch diplomat, Christiaan Huygens was an aristocratic physicist who collaborated widely with scientists and philosophers across Europe in the 17th century, including such famous names as Newton, Hooke and Descartes. Huygens' first publications were on mathematical problems, but he also studied Saturn. He was a practical scientist who patented the first pendulum clock and tried to devise a nautical clock that could be taken to sea to calculate longitude. Huygens travelled throughout Europe, especially Paris and London, meeting and working with prominent scientists on the pendulum, circular motion, mechanics and optics. Although he worked on centrifugal force alongside Newton, Huygens thought Newton's theory of gravity, with its concept of action at a distance, 'absurd'. In 1678 Huygens published his treatise on the wave theory of light.

100 metres. Its frequency, or the number of wavelengths that pass some point in one second, might be one wavelength of 100 metres in 60 seconds, or 1 cycle per minute. The fastest ocean waves are tsunami that can reach 800 kilometres per hour, the speed of a jet aircraft, slowing down to tens of kilometres per hour and rising up as they reach and swamp the coast.

To map the progress of a wave, Huygens' principle can be applied again and again as it encounters obstacles and crosses the paths of other waves. If you draw the position of a wavefront on a piece of paper, then the subsequent position can be described by using pairs of compasses to draw circles at many points along the wavefront, and drawing a smooth line through their outer edges to plot the next wave position.

The simple approach of Huygens describes waves in many circumstances. A linear wave remains straight as it propagates because the circular wavelets it produces along its length add together to form a new linear

1873

Maxwell's equations show light is an electromagnetic wave

2005

The Huygens probe lands on Titan

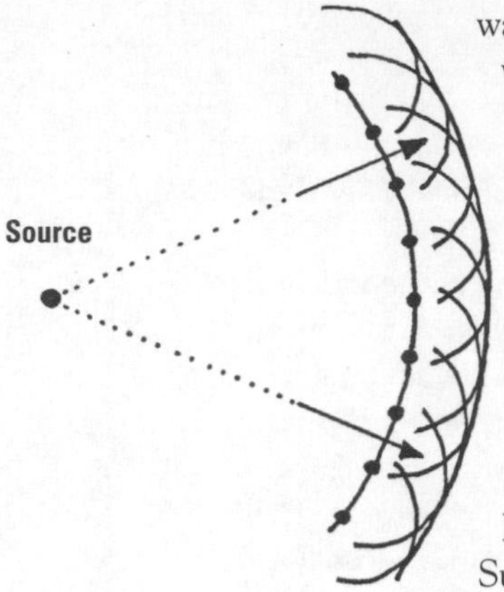

Source

wavefront ahead of the first. If you watch sets of parallel linear ocean waves as they pass through a small opening in a harbour wall, however, they distort into arcs once they pass through the gap. Only a very short length of straight wave passes through, and the arcs are formed at the edges of this unaffected remnant where, according to Huygens' principle, new circular ripples are born. If the gap is small compared with the distance between the waves then the rounded edges dominate the pattern and the transmitted wave may look almost semi-circular. This spreading out of the wave energy either side of the gap is called diffraction.

In 2004, a catastrophic tsunami created by a huge earthquake off Sumatra sped across the entire Indian Ocean. Its force in some places was diminished because the wave energy was spread out by diffraction as it travelled past and between strings of islands.

Believe your ears? Huygens' principle also explains why if you shout to someone in another room, they hear your voice as if you are standing in the doorway rather than elsewhere in the adjacent room. According to Huygens, when the waves arrive at the doorway, just like the harbour opening, a new set of point-like sources of wave energy is created there. So all the listening person knows is that these waves were generated at the doorway, the past history of the waves in the other room is lost.

Likewise, if you watch a circular ripple as it reaches the edge of a pond, its reflection produces inverted circles. The first wave point to reach the edge acts as a new source, so the backward propagation of a new circular ripple begins. Thus wave reflections can also be described using Huygens' principle.

If ocean waves move into shallower water, such as near a beach, their speed changes and the wavefronts bend inwards towards

Huygens on Titan

The Huygens space probe landed on the surface of Titan on 14 January 2005, after a seven-year journey. Contained inside a protective outer shell a few metres across, the Huygens probe carried a suite of experiments that measured the winds, atmospheric pressure, temperature and surface composition as it descended through the atmosphere to land on an icy plain. Titan is a weird world whose atmosphere and surface is damp with liquid methane. It is, some think, a place that could harbour primitive life forms such as methane-eating bacteria. Huygens was the first space probe to land on a body in the outer solar system.

the shallows. Huygens described this 'refraction' by altering the radius of the wavelets so that slower waves produced smaller wavelets. The slow wavelets do not travel as far as faster ones, so the new wavefront is at an angle to the original.

One unrealistic prediction of Huygens' principle is that if all these new wavelets are sources of wave energy then they should generate a reverse wave as well as a forward wave. So why does a wave propagate only forwards? Huygens did not have an answer and simply assumed that wave energy propagates outwards and the backwards motion is ignored. Therefore, Huygens' principle is really only a useful tool for predicting the evolution of waves rather than a fully explanatory law.

‘Each time a man stands up for an ideal . . . he sends forth a tiny ripple of hope, and crossing each other from a million different centers of energy and daring, those ripples build a current that can sweep down the mightiest walls of oppression and resistance.’

Robert Kennedy, 1966

Saturn's rings As well as wondering about ripples, Huygens also discovered Saturn's rings. He was the first to demonstrate that the planet was girdled by a flattened disk rather than flanked by extra moons or a changing equatorial bulge. He deduced that the same physics that explained the orbits of moons, Newton's gravity, would apply to many smaller bodies that would orbit in a ring. In 1655, Huygens also discovered Saturn's largest moon, Titan. Exactly 350 years later a spaceship called Cassini reached Saturn, carrying with it a small capsule, named after Huygens, which descended through the clouds of Titan's atmosphere to land on its surface of frozen methane. Titan has continents, sand dunes, lakes, and perhaps rivers, made of solid and liquid methane and ethane, rather than water. Huygens would have been amazed to think that a craft bearing his name would one day travel to that distant world, but the principle named after him can still be used to model the alien waves found there.

the condensed idea
Wave progression

16 Snell's law

Why does a drinking straw in a glass of water appear bent? It is because light travels at different speeds in air and water, causing the rays to bend. Snell's law, which describes this bending of light rays explains why puddle mirages appear over hot roads and why people in swimming pools seem to have short legs. It is being used today to help create clever materials that appear invisible.

Have you ever chuckled, watching your friend standing in a clear swimming pool, because their legs look shorter in the water than they do on land? Have you wondered why the stem of a straw looks bent as it rests on the side of your glass? Snell's law provides the answer.

When light rays cross a boundary between two materials in which light travels at different speeds, such as between air and water, the rays bend. This is called refraction. Snell's law describes how much bending occurs for transitions between different materials and is named after the 17th-century Dutch mathematician Willebrord Snellius, although he never actually published it. It is sometimes known as Snell–Descartes's law since René Descartes published a proof in 1637. This behaviour of light was well known, appearing in writings as early as the 10th century, even though it was not formalized until centuries later.

Light travels more slowly in denser materials, such as water or glass, compared with air. So a ray of sunlight travelling towards a swimming pool bends towards the pool floor when it reaches the water's surface. Because the reflected rays reach our eyes at a shallower angle, bending in reverse,

AD984 Ibn Sahl writes about refraction and lenses

1621 Snellius devises his law of refraction

1637 Descartes publishes a similar law

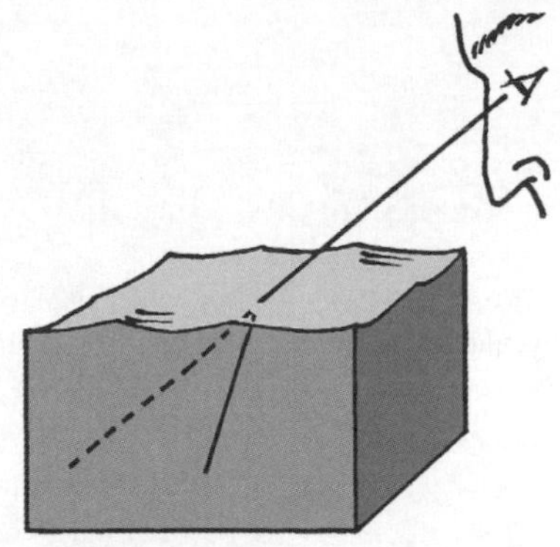

we assume that they came to us directly and so the legs of a person standing in the pool look squashed. A mirage of a puddle on a hot road is formed in a similar way. Light from the sky bends to skim the road's surface because it changes speed in the layer of hot air just above the sun-baked asphalt. Hot air is less dense than cooler air, so the light bends away from the vertical and we see the sky's reflection on the tarmac, looking like a wet puddle.

The angle by which a ray bends is related to the relative speeds at which it travels in the two materials – technically, the ratio of the speeds gives the ratio of the sine of the incident angles (measured from the vertical). So for a ray leaving air for water, and other dense substances, the ray is bent inwards and its path becomes steeper.

Refractive index Light travels at a whopping 300 million metres per second in a vacuum (e.g. empty space). The ratio of its speed in a denser material such as glass to that in a vacuum is called the refractive index of the material. A vacuum has, by definition, a refractive index of 1; something with a refractive index of 2 would slow light to half its speed in free space. A high refractive index means that light bends a lot as it passes through the substance.

Refractive index is a property of the material itself. Materials can be designed to possess specific refractive indices, which may be useful (e.g. designing lenses for glasses to correct problems with someone's vision). The power of lenses and prisms depends on their refractive index; high power lenses have high refractive indices.

Sugar sugar

Refractive index is a useful tool in wine-making and the manufacture of fruit juice. Wine makers use a refractometer to measure the concentration of sugar in the grape juice before it is turned into wine. Dissolved sugar increases the refractive index of the juice and also indicates how much alcohol it will go on to contain.

1703 Huygens publishes Snell's law

1990 Meta-materials are developed

Making a splash

Swimming pools are one of the favourite subjects of British artist David Hockney. As well as enjoying painting the optical effects of bodies gliding underwater, bathed in the bright sunshine of his California home, in 2001 Hockney caused a stir in the art world by suggesting that some famous artists used lenses to create their works as long ago as the 15th century. Simple optical devices could project a scene onto a canvas for the artist to trace and paint. Hockney has found suggestive geometries when viewing old masters, including Ingres and Caravaggio.

Refraction occurs for any waves, not just light. Ocean waves slow as the depth of the water decreases, mimicking a change in refractive index. Because of this, water waves moving at an angle to a shallow beach bend in towards the beach as they approach, which is why the surf always hits parallel to the sloping beach.

Total internal reflection Sometimes, if a light ray travelling through glass hits the boundary with air at an angle that is too shallow, then the ray will reflect back from the interface rather than continuing into the air. This is called total internal reflection, because all the light remains in the glass. The critical angle at which this happens is also determined by the relative refractive indices of the two materials. It only happens for waves travelling from a material of a high refractive index into one of lower refractive index, such as from glass to air.

Fermat's principle of least time Snell's law is a consequence of Fermat's principle of least time, which states that light rays take the quickest path through any substance. So, picking its way through a jumble of materials of various refractive indices, the light ray will choose the fastest route, favouring the low refractive index material. This is essentially a way of defining what a beam of light is, and can be derived from Huygens' principle by noting that rays that travel via the quickest path will tend to reinforce one another and create a beam, whereas light that travels off in random directions will cancel out on average. The mathematician Pierre Fermat proposed this principle in the 17th century, when the study of optics was at its height.

Meta-materials Today, physicists are designing a new class of special materials – called meta-materials – that behave in novel ways when illuminated by light or other electromagnetic waves. Meta-materials are engineered so that their appearance to light is dictated by their physical

PIERRE FERMAT 1601–65

One of the greatest mathematicians of his day, Pierre Fermat was a lawyer in Toulouse who pursued mathematics in his spare time. After writing to famous mathematicians in Paris, Fermat's reputation rose but he struggled to get anything published. He quarrelled with René Descartes about his competing theory of refraction, describing it as 'groping about in the shadows'. Descartes was angry, but Fermat proved correct. Later Fermat crystallized his work into Fermat's principle of least time, the concept that light follows the shortest path. Fermat's work was interrupted by civil war in France and outbreaks of plague. Despite false rumours that he himself had succumbed to the plague, he continued to work on number theory. He is best remembered for Fermat's last theorem that states that the sum of two cubes cannot be a cube (and so on for higher powers). Fermat wrote in the margin of a book that'I have discovered a truly remarkable proof [of this theorem] which this margin is too small to contain'. Fermat's missing proof puzzled mathematicians for three centuries before British mathematician Andrew Wiles finally proved it in 1994.

structure, rather than their chemistry. An opal is a naturally occurring meta-material – its crystal structure affects how light reflects and refracts from its surface to produce flashes of different colours.

In the late 1990s, meta-materials with negative refractive indices, in which light bends in the opposite direction at an interface, were designed. If your friend stood in a pool of liquid with a negative refractive index then, instead of seeing the front of their legs foreshortened, you would see the back of their legs projected onto their body facing you in the pool. Negative refractive index materials can be used to make 'super-lenses' that form much cleaner images than is possible with even the best glass. And in 2006, physicists succeeded in manufacturing a meta-material 'cloaking device' that appears completely invisible to microwaves.

the condensed idea
Light finds the shortest path

17 Bragg's law

The double helix structure of DNA was discovered using Bragg's law. It explains how waves travelling through an ordered solid reinforce one another to produce a pattern of bright spots whose spacing depends on the regular distances between the atoms or molecules in the solid. By measuring the emergent spot pattern the architecture of the crystalline material can be deduced.

If you are sitting in a lit room, put your hand close to the wall and you will see behind it a sharp silhouette. Move your hand further away from the wall and the shadow's outline becomes fuzzy. This is due to light diffracting around your hand. The rays of light spread inwards around your fingers as they pass by, smudging their outline. All waves behave like this. Water waves diffract around the edges of harbour walls and sound waves curve out beyond the edges of concert stages.

Diffraction can be described using Huygens' principle, which predicts the passage of a wave by considering each point on a wavefront to be a point source of further wave energy. Each point produces a circular wave, and these waves add together to describe how the wave progresses forward. If the wavefront is restricted, then the circular waves at the end points spread unimpeded. This happens when a series of parallel waves pass around an obstacle, such as your hand, or through an aperture, such as a harbour entrance or doorway.

X-ray crystallography Australian physicist William Lawrence Bragg discovered that diffraction even happens for waves travelling through crystals. A crystal is made up of many atoms stacked in a neat

timeline

AD1895 Röntgen discovers X-rays

1912 Bragg discovers his law on diffraction

WILLIAM LAWRENCE BRAGG 1890–1971

William Lawrence Bragg was born in Adelaide, where his father William Henry was a professor of mathematics and physics. Bragg junior became the first Australian to have a medical X-ray when he fell off his bicycle and broke his arm. He studied physical sciences and after graduating he followed his father to England. At Cambridge Bragg discovered his law on the diffraction of X-rays by crystals. He discussed his ideas with his father, but was upset that many people thought his father had made the discovery rather than him. During the First and Second World Wars, Bragg joined the army and worked on sonar. Afterwards, he returned to Cambridge where he set up several small research groups. In his late career, Bragg became a popular science communicator, setting up lectures for school children in London's Royal Institution and appearing regularly on television.

lattice structure, with regular rows and columns. When Bragg shone X-rays through a crystal and out onto a screen, the rays scattered off the rows of atoms. The outgoing rays piled up more in certain directions than others, gradually building up a pattern of spots. Different spot patterns appeared depending on the type of crystal used.

X-rays, discovered by German physicist Wilhelm Röntgen in 1895, were needed to see this effect because their wavelength is tiny, a thousand times less than the wavelength of visible light, and smaller than the spacing of atoms in the crystal. So X-ray wavelengths are small enough to travel through, and be strongly diffracted by, the crystal layers.

'The important thing in science is not so much to obtain new facts as to discover new ways of thinking about them.'

Sir William Bragg, 1968

The brightest X-ray spots are generated when rays traverse paths through the crystal that result in their signals being 'in phase' with one another. In-phase waves, where the peaks and

1953

X-ray crystallography is used to find the structure of DNA

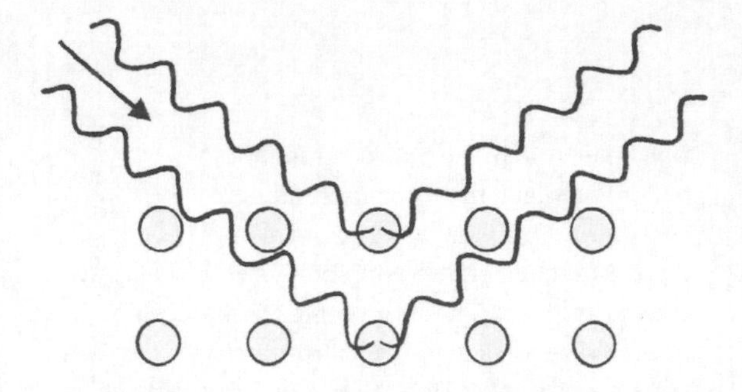

troughs are aligned, can add together to reinforce their brightness and produce spots. When 'out of phase', with the peaks and troughs misaligned, they cancel out and no light emerges. So you see a pattern of bright dots whose spacing tells you the distances between rows of atoms in the crystal. This effect of reinforcement and cancelling of waves is called 'interference'.

Bragg wrote this down mathematically by considering two waves, one reflecting from the crystal surface and the other having penetrated just one layer of atoms inside the crystal. For the second wave to be in phase and reinforce the first wave it must travel an extra distance that is a whole number of wavelengths longer than the first wave. This extra distance depends on the angle at which the rays strike and the separation between the layers of atoms. Bragg's law states how the observed interference and crystal spacings are related for a certain wavelength.

Deep structure X-ray crystallography is widely used to determine the structure of new materials and by chemists and biologists investigating

The DNA double helix

In the 1950s researchers were puzzling over the structure of DNA, one of the building blocks of life. British physicists James Watson and Francis Crick published its double helix structure in 1953, which was a major breakthrough. They acknowledged inspiration from researchers at King's College London, Maurice Wilkins and Rosalind Franklin, who had made X-ray crystallography photographs of DNA using Bragg's law. Franklin made exquisitely clear photographs showing the interference array of bright spots that ultimately gave away DNA's structure. Crick, Watson and Wilkins received the Nobel Prize for their work, but Franklin missed out because she died young. Some also think her role in the discovery was played down, perhaps due to sexist attitudes at the time. Franklin's results may also have been leaked to Watson and Crick without her awareness. Her contribution has since been acknowledged.

WILHELM RÖNTGEN 1845–1923

Wilhelm Röntgen was born in Germany's Lower Rhine, moving as a young child to the Netherlands. He studied physics at Utrecht and Zurich, and worked in many universities before his major professorships at the Universities of Würzburg and then Munich. Röntgen's work centred on heat and electromagnetism, but he is most famous for his discovery of X-rays in 1895. While passing electricity through a low-pressure gas he saw that a chemical-coated screen fluoresced even when the experiment was carried out in complete darkness. These new rays passed through many materials, including the flesh of his wife's hand placed in front of a photographic plate. He called the rays X-rays because their origin was unknown. Later, it was shown that they are electromagnetic waves like light except that they have much higher frequency.

molecular architectures. In 1953 it was used to identify the double helix structure of DNA; Francis Crick and Jim Watson famously got their idea from looking at Rosalind Franklin's X-ray interference patterns from DNA and realising that the molecules that produced them must be arranged as a double helix.

For the first time, the discovery of X-rays and crystallography techniques gave physicists tools to look at the deep structure of matter and even inside the body. Many techniques used today for medical imaging rely on similar physics. Computed tomography reassembles many X-ray slices of the body into a realistic internal picture; ultrasound maps high-frequency echoes from organs in the body; magnetic resonance imaging (MRI) scans water throughout the body's tissues identifying molecular vibrations set up using powerful magnets; and positron emission tomography (PET) follows radioactive traces as they flow through the body. So doctors and patients alike are grateful to physicists such as Bragg for developing these tools.

Braggs law is written mathematically as

$$2d \sin \theta = n\lambda$$

where d is the distance between the atomic layers, θ is the angle of incidence of the light, n is a whole number and λ is the wavelength of light.

the condensed idea
Spotting structure

18 Fraunhofer diffraction

Why can you never achieve a perfect camera image? Why is our own eyesight imperfect? Even the tiniest spot gets blurred because the light is smeared out as it passes through the eye or camera aperture. Fraunhofer diffraction describes this blurring for light rays reaching us from a distant landscape.

When you look at a faraway ship on a horizon it is impossible to read its name. You could use binoculars to do so, magnifying the picture, but why do our eyes have such limited resolution? The reason is the size of our eyes' pupils (their apertures). They need to be open wide enough to let through sufficient light to trigger the eyes' sensors, but the more open they are the more the entering light waves blur.

Light waves travelling through the lens into the eye can arrive from many directions. The wider the aperture the more directions the rays may enter from. Just as for Bragg diffraction, the different light paths interfere depending on whether their phases are aligned or misaligned. Most pass straight through in phase so forming a clear and bright central spot. But the spot's width is curtailed at the point when adjacent rays cancel each other out, and a series of dark and light alternating bands appears at the edges. It is the width of this central spot that dictates the finest detail that our eyes can pick up.

Far field Fraunhofer diffraction, named after top German lens maker Joseph von Fraunhofer, describes the blurring of images that is seen when

timeline

AD 1801
Thomas Young performs his double slit experiment

1814
Fraunhofer invents the spectroscope

light rays that fall on an aperture or lens arrive there parallel to one another. Fraunhofer diffraction, also called far-field diffraction, happens when we pass light from a distant source (e.g. the Sun or stars) through a lens. This lens could be in our eyes or in a camera or telescope. As with the limitations of eyesight, in all photography, diffraction effects smear the final image. Consequently there is a natural limit to how crisp an image can be once it has travelled through any optical system – the 'diffraction limit'. This limit scales with the wavelength of light and the reciprocal of the size of the aperture or lens. So blue images appear slightly clearer than red ones, and images taken with a larger aperture or lens will be less blurred.

Diffraction Just as the edges of your shadow hand blur due to diffraction of light around it, light spreads out when it passes through a narrow hole or aperture. Counterintuitively, the narrower the aperture the more the light spreads out. Projected onto a screen, the light that emerges from the aperture produces a central bright peak flanked by alternating dark and bright bands, or interference fringes, falling off in brightness away from the centre. Most of the rays travel straight through and reinforce, but ones that pierce at an angle interfere to produce light or dark bands.

The smaller the hole, the greater the separation between the bands, because the paths of the rays are more restricted and so more similar. If you hold two pieces of thin gauze cloth, such as silk scarves, up to the light and move them with respect to one another, similar light and dark bands arise from the overlapping threads. When they are placed on top of one another and then rotated your eye picks up a series of dark and light areas moving across the material. These interference patterns from overlapping grids are also known as 'moiré fringes'.

When the aperture or lens is circular, as is the case with our pupils and often for camera optics, the central spot and surrounding bands form a series of concentric circles called Airy rings (or Airy's disk) after the 19th-century Scottish physicist George Airy.

1822

The first Fresnel lens is used in a lighthouse

Near field Fraunhofer diffraction is often seen but if the light source is close to the aperture plane, sometimes a slightly different pattern emerges. The incident light rays are not parallel and the wavefronts arriving at the aperture are curved rather than straight. In this case a different diffraction pattern results, where the bands are no longer regularly spaced. The series of arriving wavefronts takes on the form of a set of concentric curved surfaces, like onion layers all one wavelength wide and with the light source at the centre. When these rounded wavefronts reach the aperture plane, it cuts through them like a knife cuts through the layers of an onion sliced off centre. Across the aperture, this appears as a set of rings, where each ring represents a zone where the waves that pass through it lie within a wavelength of one another.

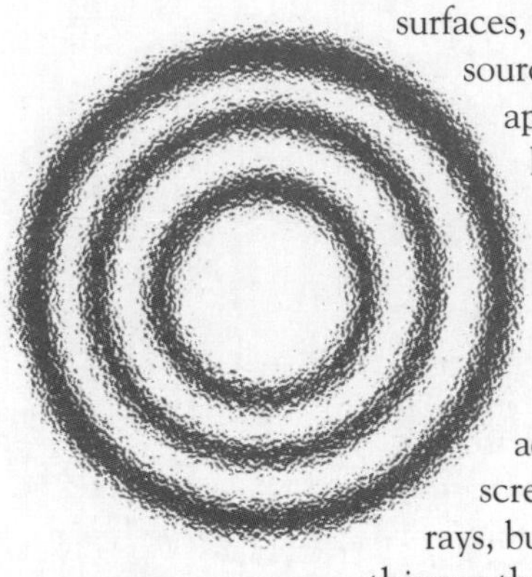

To work out how these curved rays blend with one another, you add together all the rays from the rings at the aperture. On a flat screen they display a series of dark and light bands, as for parallel rays, but the separations are no longer regular, instead becoming thinner the further from the centre you go, This is called Fresnel diffraction after Augustin Fresnel, the 19th-century French scientist who established it.

Fraunhofer diffraction

Fresnel also realized that, by editing the aperture, you could alter which phases passed through and so change the resulting pattern. He used this insight to build a new type of lens that only allowed waves in phase to pass through. One way of doing this would be to cut out a series of rings that exactly matched the positions of, say, all the negative troughs of the waves as they pass through the aperture, so only the positive peaks will come through, with hardly any interference. Alternatively, you could shift the troughs by half a wavelength and then transmit them, so that they are again in phase with the unblocked waves. Inserting rings of thicker glass at the right positions can slow the light of a particular phase by the desired amount to shift the wavelengths.

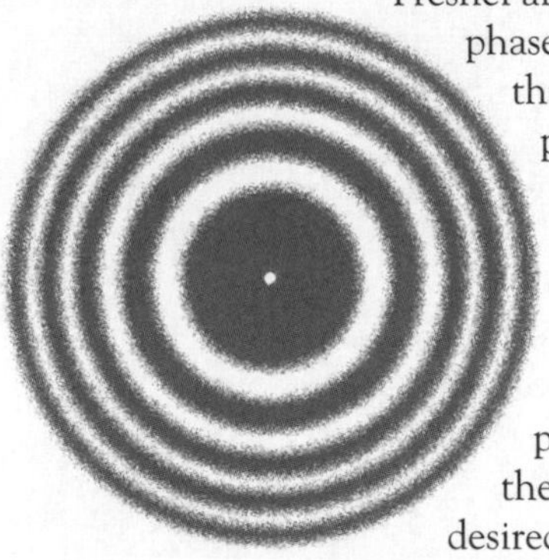

Fresnel diffraction

Fresnel himself developed lenses using this concept for lighthouses, installing the first in France in 1822. Imagine scaling up the glass lenses from a pair of spectacles to the size needed for a 50-foot lighthouse. Fresnel's alternative was a series of large but quite thin glass rings, each one

Young's double slit experiment

In his celebrated 1801 experiment, Thomas Young seemed to prove conclusively that light was a wave. When diffracting light through two slits, he not only saw a superposition of two diffraction profiles but extra stripes, due to interference from light rays that had passed through one or other of the slits. The rays interfered again to produce light and dark bands, but with a separation that corresponded to the reciprocal of the distance between the slits. So a joint pattern of fine bands against the original broad single-aperture diffraction pattern emerged. The more parallel slits were added, the sharper this second interference pattern became.

a fraction of the weight of a single convex lens. Fresnel lenses are used to focus car headlights and are sometimes stuck onto the rear windows of cars, as thin transparent plastic etched panels, to help reversing.

Gratings Fraunhofer broadened his study of interference by building the first diffraction grating. A grating has a whole series of apertures, such as many rows of parallel slits. Fraunhofer made his from aligned wires. Diffraction gratings not only spread out the light – by having multiple slits, they add further interference characteristics to the transmitted light.

Because light diffracts and interferes, it behaves in all these cases as if it were waves. But, this is not always so. Einstein and others showed that sometimes, if you look in the right way, light behaves not only as a wave but also as a particle. Quantum mechanics flowed from this observation. Astonishingly, as we will see later, in quantum versions of the double slit experiment, light knows whether to behave as a wave or as a particle and changes character just because we are watching it.

the condensed idea
Interfering light waves

19 Doppler effect

We've all heard the drop in pitch of an ambulance siren's wail as it speeds past. Waves coming from a source that is moving towards you arrive squashed together and so seem to have a higher frequency. Similarly, waves become spread out and so take longer to reach you from a source that is receding, resulting in a frequency drop. This is the Doppler effect. It has been used to measure speeding cars, blood flow and the motions of stars and galaxies in the universe.

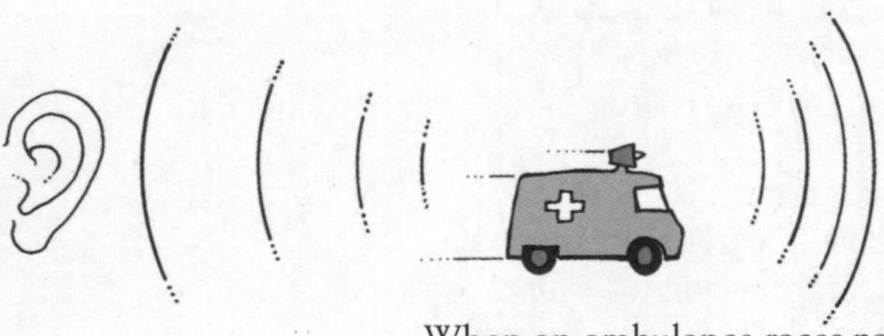

When an ambulance races past you on the street its siren wail changes in pitch, from high when it approaches to low as it recedes. This change in tone is the Doppler effect, proposed by Austrian mathematician and astronomer Christian Doppler in 1842. It arises because of the motion of the emitting vehicle relative to you, the observer. As the vehicle approaches, its sound waves pile up, the distance between each wavefront is squashed and the sound gets higher. As it speeds away, the wavefronts consistently take a little longer to reach you, the intervals get longer and the pitch drops. Sound waves are pulses of compressed air.

To and fro Imagine if someone on a moving platform, or train, was throwing balls to you continually at a frequency of one ball every three seconds, prompted by their wristwatch timer. If they are motoring towards

timeline

AD 1842
Doppler presents his paper on colour shift in starlight

CHRISTIAN DOPPLER 1803–53

Christian Doppler was born into a family of stonemasons in Salzburg, Austria. He was too frail to continue the family business and went to university in Vienna instead to study mathematics, philosophy and astronomy. Before finding a university job in Prague, Doppler had to work as a bookkeeper and even considered emigrating to America. Although promoted to professor, Doppler struggled with his teaching load, and his health suffered. One of his friends wrote 'It is hard to believe how fruitful a genius Austria has in this man. I have written to . . . many people who can save Doppler for science and not let him die under the yoke. Unfortunately I fear the worst.' Doppler eventually left Prague and moved back to Vienna. In 1842, he presented a paper describing the colour shift in the light of stars, that we now call the Doppler effect.

'It is almost to be accepted with certainty that this will in the not too distant future offer astronomers a welcome means to determine the movements and distances of such stars which, because of their unmeasurable distances from us and the consequent smallness of the parallactic angles, until this moment hardly presented the hope of such measurements and determinations.'

Although regarded as imaginative, he received a mixed reception from other prominent scientists. Doppler's detractors questioned his mathematical ability, whereas his friends thought very highly of his scientific creativity and intuition.

you it will always take a little less than three seconds for the balls to arrive because they are launched a little closer to you each time. So the rate will seem quicker to the catcher. Similarly, as the platform moves away, the balls take a little more time to arrive, travelling a little extra distance each throw, so their arrival frequency is lower. If you could measure that shift in

1912

Vesto Slipher measures red-shifts of galaxies

1992

The first detection of an extra-solar planet by the Doppler method

> **'Perhaps when distant people on other planets pick up some wavelength of ours all they hear is a continuous scream.'**
>
> **Iris Murdoch, 1919–99**

timing with your own watch then you could work out the speed of the thrower's train. The Doppler effect applies to any objects moving relative to one another. It would be the same if it was you moving on the train and the ball thrower was standing still at on a stationary platform. As a way of measuring speed, the Doppler effect has many applications. It is used in medicine to measure blood flow and also in roadside radars that catch speeding drivers.

Motion in space Doppler effects also appear frequently in astronomy, showing up wherever there is moving matter. For example, light coming from a planet orbiting a distant star would show Doppler shifts. As the planet moves towards us the frequency rises, and as it spins away its light frequency drops. Light from the approaching planet is said to be 'blue-

Extrasolar planets

More than 200 planets orbiting around stars other than our Sun have been discovered. Most are gas giants similar to Jupiter but orbiting much closer to their central stars. But a few possible rocky planets, similar to the Earth in size, have been spotted. About one in ten stars have planets, and this has fuelled speculation that some may even harbour forms of life. The great majority of planets have been found by observing the gravitational tug of the planet on its host star. Planets are tiny compared to the stars they orbit, so it is hard to see them against their star's glare. But the mass of a planet swings the star around a little, and this wobble can be seen as a Doppler shift in the frequency of a characteristic feature in the spectrum of the star.

The first extrasolar planets were detected around a pulsar in 1992 and around a normal star in 1995. Their detection is now routine but astronomers are still seeking Earth-like solar systems and trying to figure out how different planetary configurations occur. New space observatories, namely the 2006 European telescope COROT and Nasa's Kepler (in 2008), are expected to identify many Earth-like planets in the near future.

shifted'; as it moves away it has a 'red-shift'. Hundreds of planets have been spotted around distant stars since the 1990s, by finding this pattern imprinted in the glow of the central star.

Redshifts can arise not only due to planets' orbital motions, but also from the expansion of the universe itself, when it is called cosmological redshift. If the intervening space between us and a distant galaxy swells steadily as the universe expands, it is equivalent to the galaxy moving away from us with some speed. Similarly, two dots on a balloon being inflated look as if they are moving apart.

Consequently the galaxy's light is shifted to lower frequencies because the waves must travel further and further to reach us. So very distant galaxies look redder than ones nearby. Strictly speaking cosmological redshift is not a true Doppler effect because the receding galaxy is not actually moving relative to any other objects near it. The galaxy is fixed in its surroundings and it is the intervening space that is actually stretching.

To his credit, Doppler himself saw that the Doppler effect could be useful to astronomers but even he could not have foreseen how much would flow from it. He claimed to have seen it recorded in the colours of light from paired stars, but this was disputed in his day. Doppler was an imaginative and creative scientist but sometimes his enthusiasm outstripped his experimental skill. Decades later, however, redshifts were measured for galaxies by astronomer Vesto Slipher, setting the stage for the development of the big bang model of the universe. And now, the Doppler effect may help identify worlds around distant stars that could even turn out to harbour life.

the condensed idea
Perfect pitch

20 Ohm's law

Why are you safe when flying in a thunderstorm? How do lightning conductors save buildings? Why do the light bulbs in your house not dim every time you switch another one on? Ohm's law has the answers.

Electricity arises from the movement of electric charge. Electric charge is a basic property of subatomic particles that dictates how they interact with electromagnetic fields. These fields create forces that move electrically charged particles. Charge, like energy, is conserved overall; it cannot be created or destroyed but may be moved around.

Charge can be a positive or negative property. Particles of opposite charge attract one another; those with like charges repel. Electrons have a negative charge (measured by Robert Millikan in 1909) and protons a positive charge. Not all subatomic particles are charged, however. Neutrons, as the name suggests, have no charge and so are 'neutral'.

Static electricity Electricity may remain static, as a fixed distribution of charges, or flow, as an electric current. Static electricity builds up when charged particles move, so that opposite charges accumulate in different places. If you rub a plastic comb on your sleeve, for example, it becomes charged and can pick up other small items that carry opposing charge, such as small scraps of paper.

Lightning is formed in a similar way, as friction between molecules in turbulent storm clouds builds up electricity that is discharged suddenly as a lightning bolt. Lightning sparks can reach several miles in length and tens of thousands of degrees Celsius in temperature.

AD1752 Franklin conducts his lightning experiment

1826 Ohm publishes his law

BENJAMIN FRANKLIN 1706–90

Benjamin Franklin was born in Boston, USA, the 15th and youngest son of a tallow chandler. Although pushed to become a clergyman, Ben ended up working as a printer. Even after Franklin had achieved fame, he signed his letters modestly 'B. Franklin, Printer'. Franklin issued *Poor Richard's Almanac* which made him famous through memorable quotes such as 'Fish and visitors stink in three days'. Franklin was a prodigious inventor – developing the lightning rod, glass harmonica, bifocal glasses, and many more – but was fascinated most of all by electricity. In 1752 he conducted his most famous experiment, extracting sparks from a thundercloud by flying a kite in a storm. Franklin contributed to public life in his later years, introducing public libraries, hospitals and volunteer fire fighters to America and working to abolish slavery. He became a politician, conducting diplomacy between the United States, Great Britain and France during and after the American Revolution. He was a member of the Committee of Five that drafted the Declaration of Independence in 1776.

On the move Electric current, as used in the home, is a flow of charge. Metal wires conduct electricity because the electrons in metals are not tied to particular atomic nuclei and can easily be set in motion. Metals are said to be conductors of electricity. The electrons move through a metal wire like water through a pipe. In other materials, it may be positive charges that move. When chemicals are dissolved in water both electrons and positively charged nuclei (ions) float freely. Conducting materials, such as metals, allow charges to move easily through them. Materials that do not allow electricity to pass through, such as ceramics or plastics, are called insulators. Those that conduct electricity in certain circumstances only are termed semiconductors.

Like gravity, an electrical current can be created by a gradient, in this case in an electrical field or an electrical potential. So just as a change in

1909

Millikan measures the charge on a single electron

height (gravitational potential) causes a river to run downhill, a change in electrical potential between two ends of a conducting material causes a current of charge to flow through it. This 'potential difference', or voltage, drives the current flow and also gives energy to the charges.

Resistance When lightning strikes, the electric discharge flows very quickly through the ionized air to the ground. In doing so it is cancelling out the potential difference that drove it, so a lightning strike carries a huge current. It is the huge current, not the voltage, which can kill you as it surges through your body. In practice, charges cannot move at such large speeds through most materials because they encounter resistance. Resistance limits the size of current by dissipating the electrical energy as heat. To avoid being killed by lightning you could stand on an insulator, perhaps a rubber mat, which has very high resistance. Or you could hide inside a metal cage, as the lightning can flow more easily through the metal bars than your body, which, being mostly water, is not a good conductor. This construction is known as the Faraday cage, after Michael Faraday who built one in 1836. The electrical field pattern set up with a Faraday cage – a hollow conductor – means that all the charge is carried on the outside of the cage, and inside the cage it is completely neutral. Faraday cages were useful safety devices for 19th-century scientists performing with artificial lightning displays. Today they still protect electronic equipment and explain why, when you are flying through an electrical storm in a metal plane, you are safe – even if the plane scores a direct lightning hit. You are equally safe in a metal car, as long as you don't park near a tree.

In Philadelphia in 1752 Benjamin Franklin successfully 'extracted' electricity from a storm cloud with a kite.

Benjamin Franklin's lightning conductor operates in a similar way, providing a low-resistance pathway for lightning's current to follow, rather than releasing its energy into the high-resistance building that it hits. Sharp pointed rods work best because they compress the electric field onto their tip, making it more likely that the electricity will be funnelled via this route to the ground. Tall trees also concentrate the electric field, so it's not a good idea to shelter beneath one in a storm.

Lightning

It may not strike the same place twice but, on average, lightning strikes the Earth's surface a hundred times every second, or 8.6 million times a day. In the US alone, as many as 20 million lightning bolts hit the ground per year from 100,000 thunderstorms.

Circuits Electrical flows follow loops called circuits. The movement of current and energy through circuits can be described in the same way that water flows through a series of pipes. Current is similar to flow speed, voltage to the pressure of water, and resistance to the pipe width or aperture restrictions placed within it.

Georg Ohm published one of the most useful laws for interpreting circuits in 1826. Ohm's law is written algebraically as $V = IR$, which states that the voltage drop (V) is equal to the product of the current (I) and the resistance (R). According to Ohm's law, voltage is proportional to current and resistance. Double the voltage across a circuit and you double the current flowing through it if the resistance is unchanged; to maintain the same current you need a resistance twice as large. Current and resistance are inversely related, so increasing the resistance slows the current. Ohm's law applies to even quite complex circuits with many loops. The simplest circuit can be imagined as a single light bulb connected by wire to a battery. The battery supplies the potential difference needed to drive the current through the wire, and the bulb's tungsten filament provides some resistance as it converts electrical energy into light and heat. What would happen if you inserted a second light bulb into the circuit? According to Ohm's law, if the two light bulbs were placed next to one another you would have doubled the resistance and so the voltage across each of them, and thus the energy available to each, must be split in two making both bulbs glow more faintly. This wouldn't be much use if you were lighting a house – every time you plugged in another light bulb in a room they would all dim.

However, by connecting the second bulb in a linked loop directly around the first, each light bulb can be made to experience the full potential drop. The current diverts at the junction and passes through both bulbs separately before coming back together again, so the second bulb shines as brightly as the first. This sort of circuit is called a 'parallel' circuit. The former, where resistors are linked side by side is a 'series' circuit. Ohm's law can be used throughout any circuit to calculate the voltages and currents at any point.

the condensed idea
Circuit theory

21 Fleming's right hand rule

Cycling at night, you may have used a dynamo to power your bicycle lights. A corrugated rod rolls against the tyre, creating enough voltage to light two bulbs. The faster you cycle, the brighter the light glows. It works because a current has been induced in the dynamo – the direction of the current flow is given by Fleming's memorable right hand rule.

Electromagnetic induction can be used to switch between different forms of electric and magnetic fields. It is used in transformers that control the transmission of energy across the electricity grid, travel adaptors and even bicycle dynamos. When a changing magnetic field is passed over a wire coil, it produces a force on the charges inside that causes them to move and so sets up an electrical current.

> **'Faraday himself called his discovery the magnetization of light and the illumination of magnetic lines of force.'**
>
> **Pieter Zeeman, 1903**

Hidden inside the small metal canister of the dynamo are a magnet and a coil of wire. The protruding bar that spins against the wheel turns a magnet sitting inside the wire coil. Because the spinning magnet produces a changing magnetic field, charges (electrons) inside the wire are set in motion to create an electrical current. The current is said to be induced in the coil through the phenomenon of electromagnetic induction.

Rule of thumb The direction of the induced current is given by Fleming's right hand rule, named after Scottish engineer John Ambrose

timeline

AD1745 The Leyden jar capacitor is invented

1820 Ørsted links electricity and magnetism

Fleming. Hold out your right hand and point your thumb upwards, your first finger straight ahead and your second finger to the left at right angles to the forefinger. For a conductor moving upwards along your thumb, and a magnetic field pointing along the forefinger, a current will be induced in the direction of the second finger, all three mutually at right angles to one another. This handy rule is easily remembered.

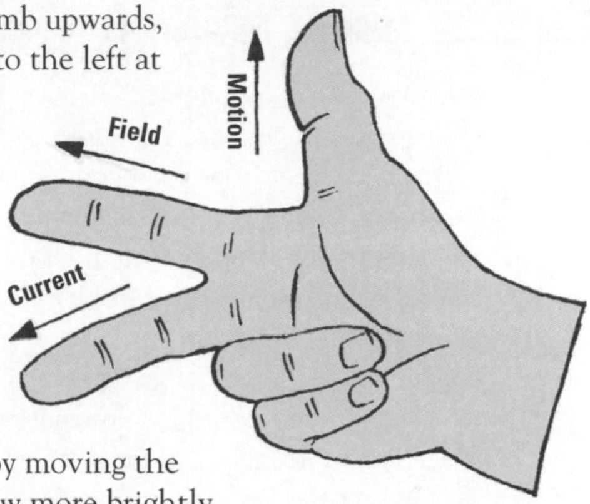

The induced current can be increased by winding the coils more tightly, so that the magnetic field changes direction more times along the length of the wire, or by moving the magnet more quickly. This is why bicycle dynamos glow more brightly when you are cycling faster. It doesn't matter if it is the magnet or the coil that is moving as long as they move relative to one another.

The relationship between the changing magnetic field and the force it induces is expressed in Faraday's law. The induced force, called the electromotive force (often abbreviated to emf), is given by the number of turns in the coil multiplied by the rate at which the magnetic flux (which increases with the magnetic field strength and area of coil) changes. The direction of the induced current always opposes that which set it up in the first place (this is known as Lenz's law). If it didn't then the whole system would self-amplify and violate the conservation of energy.

Faraday Electromagnetic induction was discovered by Michael Faraday in the 1830s. Faraday, a British physicist, was famed for his experiments with electricity. Not only did he show that magnets spin when floated in a bed of mercury, setting up the principle of the electric motor, but he also demonstrated that light is affected by magnetic fields. Rotating the plane of polarized light with a magnet, he reasoned that light must itself be electromagnetic.

1831
Faraday discovers electromagnetic induction

1873
Maxwell published his equations on electromagnetism

1892
Fleming presents transformer theory

MICHAEL FARADAY 1791–1867

British physicist Michael Faraday taught himself by reading books while an apprentice bookbinder. As a young man, Faraday attended four lectures given by the chemist Humphry Davy at the Royal Institution in London and was so impressed he wrote to Davy asking for a job. After initially being turned down, Faraday started work, spending most of his time helping others at the Royal Institution but also working on electric motors. In 1826, he started the Royal Institution's Friday evening discourses and Christmas lectures, both of which continue today. Faraday worked extensively on electricity, discovering electromagnetic induction in 1831. He became recognized as a highly skilled experimenter and was appointed to several official positions, including scientific adviser to Trinity House where he helped install electric lights in lighthouses. Perhaps surprisingly, Faraday turned down a knighthood and the presidency of the Royal Society (not once, but twice). When his health failed, Faraday spent his last days at Hampton Court in the home given to him by Prince Albert in recognition of his broad contribution to science.

Until Faraday, scientists believed that there were many different types of electricity, manifested in different situations. It was Faraday who showed that all these types could be described in a single framework based on the movement of charge. Faraday was no mathematician, and has even been called 'mathematically illiterate', but nevertheless his ideas on electric and magnetic fields were taken up by James Clerk Maxwell, another British physicist, who condensed them into his four famous equations that are still one of the bedrocks of modern physics (see page 88).

> **'Nothing is too wonderful to be true if it be consistent with the laws of nature.'**
>
> **Michael Faraday 1849**

Stored charge Faraday's name is now given to a unit of electrical charge, the Farad, which labels capacitors. Capacitors are electrical components that temporarily store charge, and are a common feature in circuits. For example, the flash unit on a disposable camera stores charge using a capacitor (while you wait for the light to come on); when you press the shutter button it releases the charge to create the flash when the picture is taken. Even using a normal battery, the voltage that builds up can be considerable, hundreds of volts, and it would give you a nasty electrical shock if you touched the capacitor.

The simplest capacitor two consists of parallel but separated metal plates with air between them. But they can be made of sandwiches of almost anything as long as the 'bread' conducts, or holds charge, and the 'filling' does not. The earliest devices for storing electrical charge in the 18th century were glass jars, called 'Leyden jars', whose interior surface was coated with metal. Today, these sandwich layers are made from materials such as aluminium foil, niobium, paper, polyester and Teflon. If a capacitor is connected to a battery, when the battery is switched on, opposite charges build up on each plate. When the battery is switched off, the charges are released as a current. The current decays because its 'pressure' drops off as the charge difference reduces. Because it takes time to charge up and discharge capacitors, they can substantially delay the flow of charge around circuits. Capacitors are often used together with inductors (such as coils of wire that may add induced currents) to build circuits where the charge oscillates back and forth.

Transformers Electromagnetic induction isn't only used in dynamos and motors but also in electrical transformers. A transformer works by first generating a changing magnetic field and then using that field to induce a second current in a nearby coil. A simple transformer consists of a magnetic ring with two separated wire coils wrapped around it. A changing electric field is pumped through the first coil, setting up an oscillating magnetic field throughout the magnet. This changing field then induces a new current in the second coil.

By Faraday's law, the size of the induced current depends on the number of loops in the coil, so the transformer can be designed to tune the size of the output current. When electricity is sent around a national grid it is more efficient and safer to send it as low-current, high-voltage electricity. Transformers are used at both ends of the network, ramping up the voltage to lower the current for distribution and lowering it at the household end. As you will know if you have ever touched the brick of a computer power adaptor or a travel adaptor, transformers are not 100% efficient as they heat up and often hum, losing energy to sound, vibration and heat.

the condensed idea
Induction rules

22 Maxwell's equations

Maxwell's four equations are a cornerstone of modern physics and the most important advance since the universal theory of gravitation. They describe how electric and magnetic fields are two sides of the same coin. Both types of field are manifestations of the same phenomenon – the electromagnetic wave.

Early 19th-century experimenters saw that electricity and magnetism could be changed from one form into the other. But James Clerk Maxwell completed one of the major achievements of modern physics when he managed to describe the whole field of eletromagnetism in just four equations.

Electromagnetic waves Electric and magnetic forces act on charged particles and magnets. Changing electric fields generate magnetic fields, and vice versa. Maxwell explained how both arise from the same phenomenon, an electromagnetic wave, which exhibits both electric and magnetic characteristics. Electromagnetic waves contain a varying electric field, accompanied by a magnetic field that varies similarly but lies at right angles to the other.

Maxwell measured the speed of electromagnetic waves travelling through a vacuum, showing it to be essentially the same as the speed of light. Combined with the work of Hans Christian Ørsted and Faraday, this confirmed that light was also a propagating electromagnetic disturbance. Maxwell showed that light waves, and all electromagnetic waves, travel at

timeline

AD 1600
William Gilbert investigates electricity and magnetism

1752
Benjamin Franklin conducts his experiments on lightning

1820
Ørsted links electricity and magnetism

a constant speed in a vacuum of 300 million metres per second. This speed is fixed by the absolute electric and magnetic properties of free space.

> **'We can scarcely avoid the conclusion that light consists in the transverse undulations of the same medium which is the cause of electric and magnetic phenomena.'**
>
> **James Clerk Maxwell, c.1862**

Electromagnetic waves can have a range of wavelengths and cover a whole spectrum beyond visible light. Radio waves have the longest wavelengths (metres or even kilometres), visible light has wavelengths that are similar to the spacing between atoms, while at the highest frequencies are X-rays and gamma rays. Electromagnetic waves are used mainly for communications, via the transmission of radio waves, television and mobile phone signals. They can provide heat energy, such as in microwave ovens, and are often used as probes (e.g. medical X-rays and in electronic microscopes).

The electromagnetic force exerted by electromagnetic fields is one of the four fundamental forces, along with gravity and the strong and weak nuclear forces, that hold atoms and nuclei together. Electromagnetic forces are crucial in chemistry where they bind charged ions together to form chemical compounds and molecules.

Fields Maxwell started out by trying to understand Faraday's work describing electric and magnetic fields experimentally. In physics, fields are the way in which forces are transmitted across distances. Gravity operates across even the vast distances of space, where it is said to produce a gravitational field. Similarly, electric and magnetic fields can affect charged particles quite far away. If you have played with iron filings sprinkled over a sheet of paper with a magnet below it, you will have seen that the magnetic force moves the iron dust into looped contours stretching from the north to the south pole of the magnet. The magnet's strength also falls off as you move further away from it. Faraday had mapped these 'field lines' and worked out simple rules. He also mapped similar field lines for electrically charged shapes but was not a trained

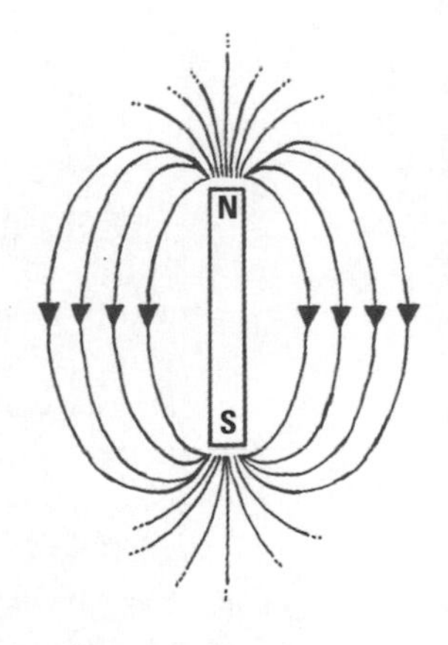

1831
Faraday discovers electromagnetic induction

1873
Maxwell publishes his equations on electromagnetism

1905
Einstein publishes the special theory of relativity

mathematician. So it fell to Maxwell to try to unite these various ideas into a mathematical theory.

Four equations To every scientist's surprise, Maxwell succeeded in describing all the various electromagnetic phenomena in just four fundamental equations. These equations are now so famous that they don some T-shirts followed by the comment 'and so god created light'. Although we now think of electromagnetism as one and the same thing, at the time this idea was radical, and as important as if we united quantum physics and gravity today.

$$\nabla \cdot D = \rho$$
$$\nabla \times H = J + (\delta D/\delta t)$$
$$\nabla \cdot B = 0$$
$$\nabla \times E = -(\delta B/\delta t)$$

Maxwell's equations

The first of Maxwell's equations is Gauss's law, named after 19th-century physicist Carl Friedrich Gauss, which describes the shape and strength of the electric field generated by a charged object. Gauss's law is an inverse square law, mathematically similar to Newton's law of gravity. Like gravity, electric field drops off away from the surface of a charged object in proportion to the square of the distance. So the field is four times weaker if you move twice as far away from it.

Although there is no scientific evidence that mobile phone signals are bad for your health, the inverse square law explains why it might be safer to have a mobile phone mast close to your home rather than far away. The

JAMES CLERK MAXWELL 1831–79

James Clerk Maxwell was born in Edinburgh, Scotland. He grew up in the countryside where he became curious about the natural world. After his mother died, he was sent to school in Edinburgh where he was given the nickname 'dafty', because he was so absorbed in his studies. As a student at Edinburgh University and later at Cambridge, Maxwell was thought clever if disorganized. After graduation, he extended Michael Faraday's work on electricity and magnetism and condensed it into equations. Maxwell moved back to Scotland when his father became ill and tried to get a job again at Edinburgh. Losing out to his old mentor, he went to Kings College London where he carried out his most famous work. Around 1862 he calculated that the speed of electromagnetic waves and light were the same and 11 years later he published his four equations of electromagnetism.

field from the transmitter mast drops off rapidly with distance, so is very weak by the time it reaches you. In comparison, the field from the mobile phone is strong because it is held so close to your head. So, the closer the mast the less power the potentially more dangerous phone uses when you talk on it. Nevertheless, people are often irrational and fear masts more.

> **'Any intelligent fool can make things bigger and more complex . . . It takes a touch of genius – and a lot of courage to move in the opposite direction.'**
>
> attributed to Albert Einstein, 1879–1955

The second of Maxwell's equations describes the shape and strength of the magnetic field, or the pattern of the magnetic field lines, around a magnet. It states that the field lines are always closed loops, from the north to the south pole. In other words all magnets must have both a north and south pole – there are no magnetic monopoles and a magnetic field always has a beginning and an end. This follows from atomic theory where even atoms can possess magnetic fields and grand-scale magnetism results if these fields are all aligned. If you chop a bar magnet in half, you always reproduce the north and south poles on each half. No matter how much you divide the magnet, the smaller shards retain both poles.

The third and fourth equations are similar to one another and describe electromagnetic induction. The third equation tells how changing currents produce magnetic fields, and the fourth how changing magnetic fields produce electric currents. The latter is otherwise familiar as Faraday's law of induction.

Trying to meld electromagnetism and quantum theory in the 1930s, British physicist Paul Dirac predicted that magnetic monopoles may exist. None has yet been seen to verify this idea.

Describing so many phenomena in such a few simple equations was a major feat that led to Einstein rating Maxwell's achievement on a par with that of Newton. Einstein took Maxwell's ideas and incorporated them further into his relativity theories. In Einstein's equations, magnetism and electricity were manifestations of the same thing seen by viewers in different frames of reference; an electric field in one moving frame would be seen as a magnetic field in another. Perhaps it was Einstein then who ultimately contrived that electric and magnetic fields are truly one and the same thing.

the condensed idea
. . . and so there was light

23 Planck's law

Why do we say a fire is red hot? And why does steel glow first red, then yellow, then white, when it is heated? Max Planck described these colour changes by knitting together the physics of heat and light. Describing light statistically rather than as a continuous wave, Planck's revolutionary idea seeded the birth of quantum physics.

In a famous 1963 speech, British Prime Minister Harold Wilson marvelled at 'the white heat of this [technological] revolution'. But where does this phrase 'white heat' come from?

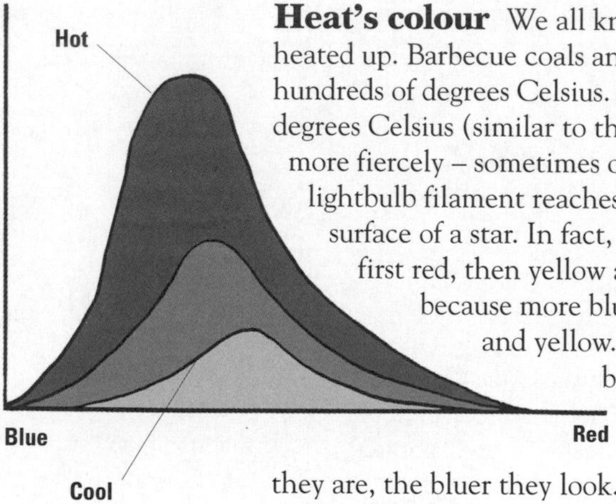

Heat's colour We all know that many things glow when they are heated up. Barbecue coals and electric stove rings turn red, reaching hundreds of degrees Celsius. Volcanic lava, approaching a thousand degrees Celsius (similar to the temperature of molten steel), can glow more fiercely – sometimes orange, yellow or even white hot. A tungsten lightbulb filament reaches over 3000 degrees Celsius, similar to the surface of a star. In fact, with increasing temperature, hot bodies glow first red, then yellow and eventually white. The light looks white because more blue light has been added to the existing red and yellow. This spread of colours is described as a black-body curve.

Stars also follow this sequence: the hotter they are, the bluer they look. The Sun, at 6000 kelvins, is yellow, while the surface of the red giant Betelgeuse (found in Orion) has a temperature of only half that. Hotter stars such as Sirius, the brightest star in the sky, whose scorching surface reaches 30,000 kelvins, look blue-white. As the temperatures increase, more and more high-frequency blue light is given

timeline

AD 1862 The term 'black body' is used by Gustav Kirchhoff

1901 Planck publishes his law of black-body radiation

MAX PLANCK 1858–1947

Max Planck was schooled in Munich, Germany. Hoping for a career in music, he sought advice on what to study from a musician, but was told if he needed to ask the question he should study something else. His physics professor was no more encouraging, telling him physics as a science was complete and nothing more could be learned. Luckily Planck ignored him and continued his research, instigating the concept of quanta. Later in Planck's life he suffered the deaths of his wife and several children, including two sons killed in the world wars. Nevertheless, Planck remained in Germany and tried to rebuild physics research there following the wars. Today, many prestigious Max Planck research institutes are named after him.

off. In fact, the strongest light from hot stars is so blue that most of it radiates in the ultraviolet part of the spectrum.

Black-body radiation Nineteenth-century physicists were surprised to find that the light emitted when objects were heated followed the same pattern, irrespective of the substance they tested. Most of the light was given off at one particular frequency. When the temperature was raised, the peak frequency shifted to bluer (shorter) wavelengths, moving from red through yellow to blue–white.

We use the term black-body radiation for a good reason. Dark materials are best able to radiate or absorb heat. If you've worn a black T-shirt on a hot day you'll know it heats up in the sun more than a white one. White reflects sunlight better, which is why houses in hot climates are often painted white. Snow reflects sunlight too. Climate scientists worry that the Earth will heat up more rapidly should the polar ice caps melt and reflect less sunlight back out into space. Black

> **'[the black-body theory was] an act of despair because a theoretical interpretation had to be found at any price, no matter how high that might be.'**
>
> **Max Planck, 1901**

1905
Einstein identifies the photon, and the ultraviolet catastrophe is disproved

1996
COBE satellite data determines the precise temperature of the cosmic microwave background radiation

Planck's legacy in space

The most perfect black-body spectrum hails from a cosmic source. The sky is bathed in the faint glow of microwaves that are the afterglow of the fireball of the big bang itself, redshifted by the expansion of the universe to lower frequencies. This glow is called the cosmic microwave background radiation. In the 1990s, NASA's COBE satellite (COsmic Background Explorer) measured the temperature of this light – it has a black-body spectrum of 2.73 K, and is so uniform that this is still the purest black-body curve measured. No material on Earth has such a precise temperature. The European Space Agency recently honoured Planck by naming their new satellite after him. It will map the cosmic microwave background in great detail.

objects not only absorb but also release heat more quickly than white ones. This is why the surfaces of stoves or hearths are painted black – not just to hide the soot!

A revolution Although physicists had measured the black-body graphs, they could not fathom them or explain why the frequency peaked at a single colour. Leading thinkers Wilhelm Wien, Lord Rayleigh and James Jeans worked out partial solutions. Wien described the dimming at bluer frequencies mathematically, while Rayleigh and Jeans explained the rising red spectrum, but both formulae failed at the opposite ends. Rayleigh and Jeans's solution, in particular, raised problems because it predicted that an infinite amount of energy would be released at ultraviolet wavelengths and above, due to the ever rising spectrum. This obvious problem was dubbed the 'ultraviolet catastrophe'.

In trying to understand black-body radiation, German physicist Max Planck joined the physics of heat and light together. Planck was a physics purist who liked returning to basics to derive physical principles. He was fascinated by the concept of entropy and the second law of thermodynamics. He considered this and Maxwell's equations to be fundamental laws of nature and set about proving how they were linked. Planck had complete faith in mathematics – if his equations told him something was true, it didn't matter if everyone else thought differently.

Planck reluctantly applied a clever fix to make his equations work. His insight was to treat electromagnetic radiation in the same way as thermodynamics experts treated heat. Just as temperature is the sharing of heat energy amongst many particles, Planck described light by allocating electromagnetic energy among a set of electromagnetic oscillators, or tiny subatomic units of electromagnetic field.

To fix the mathematics, Planck scaled the energy of each electromagnetic unit with frequency, such that $E = h\nu$, where E is energy, ν is light frequency, and h is a constant scaling factor now known as Planck's constant. Theses units were called 'quanta', from the Latin for 'how much'.

In the new picture of energy quanta, the high-frequency electromagnetic oscillators each took on high energy. So, you couldn't have many of them in any system without blowing the energy limit. Likewise, if you received your monthly salary in 100 bank notes of mixed denominations, you'd receive mostly medium denominations plus a few of higher and lower value. By working out the most probable way of sharing electromagnetic energy between the many oscillators, Planck's model put most of the energy in the middle frequencies – it fitted the peaked black-body spectrum. In 1901, Planck published this law, linking light waves with probability, to great acclaim. And it was soon seen that his new idea solved the 'ultraviolet catastrophe' problem.

Planck's quanta were just a construction for working out the mathematics of his law; he didn't for a moment imagine his oscillators were real. But, at a time when atomic physics was developing fast, Planck's novel formulation had surprising implications. Planck had planted a seed that would grow to become one of the most important areas of modern physics: quantum theory.

the condensed idea
Energy budget

24 Photoelectric effect

When ultraviolet light shines on a copper plate, electricity is produced. This 'photoelectric' effect remained a mystery until Albert Einstein, inspired by Max Planck's use of energy quanta, concocted the idea of the light particle, or photon. Einstein showed how light could behave as a stream of photon pellets as well as a continuous wave.

The dawn of the 20th century opened a new window onto physics. In the 19th century it was well known that ultraviolet light mobilized electrons to produce currents in a metal; understanding this phenomenon led physicists to invent a whole new language.

Blue batters The photoelectric effect generates electric currents in metals when they are illuminated by blue or ultraviolet light, but not red light. Even a bright beam of red light fails to trigger a current. Charge flows only when the light's frequency exceeds some threshold, which varies for different metals. The threshold indicated that a certain amount of energy needed to be built up before the charges could be dislodged. The energy to free them must come from the light but, at the end of the 19th century, the mechanism by which this happened was unknown. Electromagnetic waves and moving charges seemed to be very different physical phenomena, and uniting them was a major puzzle.

> 'There are two sides to every question.'
> **Protagoras, 485–421 BC**

Photons In 1905, Albert Einstein came up with a radical idea to explain the photoelectric effect. It was this work, rather than relativity, that won him the Nobel Prize in 1921. Inspired by Max

timeline

AD 1839
Alexandre Becquerel observes the photoelectric effect

1887
Hertz measures sparks across gaps caused by ultraviolet light

1899
J.J. Thomson confirms electrons are generated by the incident light

Planck's earlier use of quanta to budget the energy of hot atoms, Einstein imagined that light too could exist in little energy packets. Einstein borrowed wholesale Planck's mathematical definition of quanta, the proportionality of energy and frequency linked by Planck's constant, but applied it to light rather than atoms. Einstein's light quanta were later named photons. Photons have no mass and travel at the speed of light.

Rather than bathing the metal with continuous light waves, Einstein suggested that individual photon bullets hit electrons in the metal into motion to produce the photoelectric effect. Because each photon carries a certain energy, scaling with its own frequency, the bumped electron's energy also scales with the light's frequency. A photon of red light (with a low frequency) cannot carry enough energy to dislodge an electron, but a blue photon (light with a higher frequency) has more energy and can set it rolling. An ultraviolet photon has more energy still, so it can slam into an electron and donate even more speed. Turning up the brightness of light changes nothing, it doesn't matter that you have more red photons if each is incapable of shifting the electrons. It's like firing ping pong balls at a weighty sports utility vehicle. Einstein's idea of light quanta was unpopular at first, because it opposed the wave description of light summarized in Maxwell's equations that most physicists revered. However, the climate altered when experiments showed Einstein's wacky idea to be true. They confirmed the energies of the liberated electrons scaled proportionally with the frequency of light.

Wave–particle duality Einstein's proposal was not only controversial but it raised the uncomfortable idea that light was both a wave and a particle, called wave particle duality. Light's behaviour up until Maxwell wrote down his equations had always followed that of a wave, bending round obstacles, diffracting, reflecting and interfering. Here, Einstein really rocked the boat by showing that light was also a stream of photon torpedoes.

1901
Planck introduces the concept of energy quanta

1905
Einstein proposes the theory of light quanta

1924
de Broglie proposes that particles can behave as waves

ALBERT EINSTEIN 1879–1955

1905 was an annus mirabilis for a part-time German-born physicist working as a clerk in the Swiss Patent Office. Albert Einstein published three physics papers in the German journal, *Annalen der Physik*. They explained Brownian motion, the photoelectric effect and special relativity, and each one was groundbreaking work. Einstein's reputation grew until, in 1915, he produced his theory of general relativity, confirming him as one of the greatest scientists of all time. Four years later, observations made during a solar eclipse verified his general relativity theory and he became world famous. In 1921 Einstein was awarded the Nobel Prize for his work on the photoelectric effect, which influenced the development of quantum mechanics.

'The body's surface layer is penetrated by energy quanta whose energy is converted at least partially into kinetic energy of the electrons. The simplest conception is that a light quantum transfers its entire energy to a single electron.'

Albert Einstein, 1905

Physicists are still struggling with this tension. Today, we even know that light seems to know whether to behave as one or the other under different circumstances. If you set up an experiment to measure its wave properties, such as passing it through a diffraction grating, it behaves as a wave. If instead you try to measure its particle properties it is similarly obliging.

Physicists have tried to devise clever experiments to catch light out, and perhaps reveal its true nature, but so far they have all failed. Many are variants of Young's double-slit experiment but with components that can be switched in and out. Imagine a light source whose rays pass through two narrow slits onto a screen. With both slits open you see the familiar dark and light stripes of interference fringes. So light, as we know, is a wave. However, by dimming the light enough, at some point the level becomes so low that individual photons pass through the apparatus one by one, and a detector can catch the flashes as they arrive at the screen. Even if you do this, the photons continue to pile up into the striped interference pattern.

So, how does a single photon know whether to go through one or the other slit to build up the interference pattern? If you're quick, you could close one of the slits after the photon has left the light source, or even after it has travelled through the slits but before it hits the screen. In every case physicists have been able to test, the photons know whether there were one or two slits present at the time they went through. And even though only single photons are flying across, it appears as if each photon goes through both slits simultaneously.

Solar cells

The photoelectric effect is used today in solar panels where light liberates electrons, usually from semi-conductor materials like silicon rather than pure metals.

Put a detector in one of the slits (so you know whether the photon went through that one or the other) and strangely the interference pattern disappears – you're left with a simple pile up of photons on the screen and no interference stripes. So no matter how you try to catch them out, photons know how to act. And they act as both waves and particles, not one or the other.

Matter waves In 1924, Louis-Victor de Broglie suggested the converse idea that particles of matter could also behave as waves. He proposed that all bodies have an associated wavelength, implying that particle–wave duality was universal. Three years later the matter-wave idea was confirmed when electrons were seen to diffract and interfere just like light. Physicists have now also seen larger particles behaving like waves, such as neutrons, protons and recently even molecules including microscopic carbon footballs or 'bucky balls'. Bigger objects, like ball bearings and badgers, have minuscule wavelengths, too small to see, so we cannot spot them behaving like waves. A tennis ball flying across a court has a wavelength of 10^{-34} metres, much smaller than a proton's width (10^{-15} m).

As we have seen that light is also a particle and electrons are sometimes waves, the photoelectric effect has come full circle.

the condensed idea
Photon bullets

25 Schrödinger's wave equation

How can we say where a particle is if it is also spread out as a wave? Erwin Schrödinger wrote down a landmark equation that describes the probability of a particle being in some location while behaving as a wave. His equation went on to illustrate the energy levels of electrons in atoms, launching modern chemistry as well as quantum mechanics.

According to Einstein and Louis-Victor de Broglie, particles and waves are closely entwined. Electromagnetic waves, including light, take on both characteristics and even molecules and subatomic particles of matter can diffract and interfere as waves.

But waves are continuous, and particles are not. So how can you say where a particle is if it is spread out in the form of a wave? Schrödinger's equation, devised by Austrian physicist Erwin Schrödinger in 1926, describes the likelihood that a particle that is behaving as a wave is in a certain place, using the physics of waves and probability. It is one of the cornerstones of quantum mechanics, the physics of the atomic world.

Schrödinger's equation was first used to describe the positions of electrons in an atom. Schrödinger tried to describe electrons' wave-like behaviour and also incorporated the concept of energy quanta introduced by Max Planck, the idea that wave energy comes in basic building blocks whose energy scales with wave frequency. Quanta are the smallest blocks, giving a fundamental graininess to any wave.

timeline

AD 1897

J.J. Thomson discovers the electron

Bohr's atom It was Danish physicist Niels Bohr who applied the idea of quantized energy to electrons in an atom. Because electrons are easily liberated from atoms, and negatively charged, Bohr thought that, like planets in orbit around the Sun, electrons were held in orbit about a positively charged nucleus. However, electrons could exist only with certain energies, corresponding to multiples of basic energy quanta. For electrons held within an atom, these energy states should restrict the electrons to distinct layers (or 'shells') according to energy. It is as if the planets could only inhabit certain orbits, defined by energy rules.

Bohr's model was very successful, especially in explaining the simple hydrogen atom. Hydrogen contains just one electron orbiting around a single proton, a positively charged particle that acts as the nucleus. Bohr's hierarchy of quantized energies explained conceptually the characteristic wavelengths of light that were emitted and absorbed by hydrogen.

Just like climbing a ladder, if the electron in a hydrogen atom is given an energy boost, it can jump up to a higher rung, or shell. To hop up to the higher rung the electron must absorb energy from a photon with exactly the right energy to do it. So a particular frequency of light is needed to raise the electron's energy level. Any other frequency will not work. Alternatively, once boosted, the electron could jump back down to the rung below, emitting a photon of light of that frequency as it does so.

Spectral fingerprints Moving electrons up the energy ladder, hydrogen gas can absorb a series of photons of characteristic frequencies corresponding to the energy gaps between rungs. If white light is shone through the gas, these frequencies appear blacked out because all the light at each gap frequency is absorbed. Bright lines result instead if the hydrogen is hot and its electrons started out high on the ladder. These characteristic energies for hydrogen can be measured, and they agree with the

1913
Bohr proposes electrons orbit an atomic nucleus

1926
Schrödinger devises his wave equation

predictions of Bohr. All atoms produce similar lines, at different characteristic energies. So they are like fingerprints that can identify individual chemical species.

Wave functions Bohr's energy levels worked well for hydrogen, but less well for other atoms with more than one electron and with heavier nuclei. Moreover, there was still de Broglie's conundrum that electrons should also be thought of as waves. So each electron orbit could equally be considered a wavefront. But, thinking of it as a wave meant it was impossible to say where the electron was at any time.

Schrödinger, inspired by de Broglie, wrote down an equation that could describe the position of a particle when it was behaving as a wave. He was only able to do this statistically by incorporating probability. Schrödinger's important equation is a fundamental part of quantum mechanics.

Boxed in

A lone particle floating in free space has a wave function that looks like a sine wave. If it is trapped inside a box, then its wave function must drop to zero at the box walls, and outside, because it cannot be there. The wave function inside the box can be determined by considering the allowed energy levels, or energy quanta, of the particle, which must always be greater than zero. Because only specific energy levels are allowed by quantum theory, the particle will be more likely to be in some places than others and there are places within the box where the particle would never be found, where the wave function is zero. More complicated systems have wave functions that are a combination of many sine waves and other mathematical functions, like a musical tone made up of many harmonics. In conventional physics, we would use Newton's laws to describe the motion of a particle in a box (such as a miniature ball bearing). At any instant, we would know exactly where it is and the direction in which it is moving. In quantum mechanics, however, we can only talk about the probability of the particle being in some place at some time and, because energy quantization seeps in on atomic scales, there are favoured places where the particle will be found. But we cannot say exactly where it is, because it is also a wave.

> **God runs electromagnetics by wave theory on Monday, Wednesday, and Friday, and the Devil runs them by quantum theory on Tuesday, Thursday, and Saturday.**
>
> **Sir William Bragg, 1862–1942**

Schrödinger introduced the idea of a wave function to express the probability of the particle being in a given place at some time, and to include all the knowable information about that particle. Wave functions are notoriously difficult to comprehend as we do not witness them in our own experience and find it very hard to visualize and even interpret them philosophically.

The breakthrough that Schrödinger's equation heralded also led to models of electron orbitals in atoms. These are probability contours, outlining regions where electrons are 80–90% likely to be located (raising the issue that with some small probability they could be somewhere else entirely). These contours turned out not to be spherical shells, as envisaged by Bohr, but rather more stretched shapes, such as dumb-bells or donuts. Chemists now use this knowledge to engineer molecules.

Schrödinger's equation revolutionized physics by bringing the wave–particle duality idea not only to atoms but to all matter. Together with Werner Heisenberg and others, Schrödinger truly is one of the founding fathers of quantum mechanics.

the condensed idea
Here, there, but not everywhere

26 Heisenberg's uncertainty principle

Heisenberg's uncertainty principle states that the speed (or momentum) and position of a particle at an instant cannot both be known exactly – the more precisely you measure one, the less you can find out about the other. Werner Heisenberg argued that the very act of observing a particle changes it, making precise knowledge impossible. So neither the past nor the future behaviour of any subatomic particle can be predicted with certainty. Determinism is dead.

In 1927, Heisenberg realized that quantum theory contained some strange predictions. It implied that experiments could never be done in complete isolation because the very act of measurement affected the outcome. He expressed this connection in his 'uncertainty principle' – you cannot simultaneously measure both the position and momentum of a subatomic particle (or equivalently its energy at an accurate time). If you know one then the other is always uncertain. You can measure both within certain bounds, but the more tightly these bounds are specified for one, the looser they become for the other. This uncertainty, he argued, was a deep consequence of quantum mechanics – it had nothing to do with a lack of skill or accuracy in measuring.

AD 1687

Newton's laws of motion imply a deterministic universe

Uncertainty In any measurement, there is an element of uncertainty in the answer. If you measure the length of a table with a tape measure, you can say it is one metre long but the tape can only say so to within one millimetre because that is the size of the smallest tick mark on it. So the table could really be 99.9 centimetres or 100.1 centimetres long and you wouldn't know.

It is easy to think of uncertainties as being due to the limitations of your measuring device, such as the tape, but Heisenberg's statement is profoundly different. It states that you can never know both quantities, momentum and position, exactly at the same time no matter how accurate an instrument you use. It is as if when you measure a swimmer's position you cannot know her speed at the same instant. You can know both roughly, but as soon as you tie one down the other becomes more uncertain.

Measurement How does this problem arise? Heisenberg imagined an experiment that measured the motion of a subatomic particle such as a neutron. A radar could be used to track the particle, by bouncing electromagnetic waves off it. For maximum accuracy you would choose gamma rays, which have very small wavelengths. However, because of wave–particle duality the gamma ray beam hitting the neutron would act like a series of photon bullets. Gamma rays have very high frequencies and so each photon would carry a great deal of energy. As a hefty photon hit the neutron, it would give it a big kick that would alter its speed. So, even if you knew the position of the neutron at that instant, its speed changes unpredictably because of the very process of observation.

If you used softer photons with lower energies, to minimize the velocity change, then their wavelengths would be longer and so the accuracy with which you could measure the position would now be degraded. No matter how you optimize the experiment, you cannot learn both the particle's position and speed simultaneously. There is a fundamental limit expressed in Heisenberg's uncertainty principle.

1901
Planck's law uses statistical techniques

1927
Heisenberg publishes his uncertainty principle

WERNER HEISENBERG 1901–76

Werner Heisenberg lived in Germany through two world wars. An adolescent during the First World War, Heisenberg joined the militarized German youth movement that encouraged structured outdoor and physical pursuits. Heisenberg worked on farms in the summer, using the time to study mathematics. He studied theoretical physics at Munich University, finding it hard to shuttle between his love of the countryside and the abstract world of science. After his doctorate, Heisenberg took up academic posts, and met Einstein on a visit to Copenhagen. In 1925, Heisenberg invented the first form of quantum mechanics, known as matrix mechanics, receiving the Nobel Prize for this work in 1932. Nowadays, he is best known for the uncertainty principle, formulated in 1927.

During the Second World War Heisenberg headed the unsuccessful German nuclear weapons project, and worked on a nuclear fission reactor. It is debatable whether the German inability to build a nuclear weapon was deliberate or simply due to lack of resources. After the war he was arrested by the Allies and interned with other German scientists in England before returning to research in Germany afterwards.

In reality, what is going on is more difficult to comprehend, because of the coupled wave–particle behaviour of both subatomic particles and electromagnetic waves. The definitions of particle position, momentum, energy and time are all probabilistic. Schrödinger's equation describes the probability of a particle being in a certain place or having a certain energy according to quantum theory, as embodied in the wave function of the particle that describes all its properties.

$$\Delta x \Delta p > \frac{\hbar}{2}$$

$$\Delta E \Delta t > \frac{\hbar}{2}$$

Heisenberg's uncertainty principle

Heisenberg was working on quantum theory at about the same time as Schrödinger. Schrödinger preferred to work on the wave-like aspects of subatomic systems, whereas Heisenberg investigated the stepped nature of the energies. Both physicists developed ways of describing quantum systems mathematically according to their own biases; Schrödinger using the mathematics of waves and Heisenberg using matrices, or two-dimensional tables of numbers, as a way of writing down the sets of properties.

The matrix and wave interpretations both had their followers, and both camps thought the other group was wrong. Eventually they pooled their

> **'The more precisely the position is determined, the less precisely the momentum is known in this instant, and vice versa.'**
>
> **Werner Heisenberg, 1927**

resources and came up with a joint description of quantum theory that became known as quantum mechanics. It was while trying to formulate these equations that Heisenberg spotted uncertainties that could not go away. He brought these to the attention of a colleague, Wolfgang Pauli, in a letter in 1927.

Indeterminism The profound implications of the uncertainty principle were not lost on Heisenberg and he pointed out how it challenged conventional physics. First, it implied that the past behaviour of a subatomic particle was not constrained until a measurement of it was made. According to Heisenberg "the path comes into existence only when we observe it". We have no way of knowing where something is until we measure it. He also noted that the future path of a particle cannot be predicted either. Because of these deep uncertainties about its position and speed, the future outcome was also unpredictable.

Both of these statements caused a major rift with the Newtonian physics of the time, which assumed that the external world existed independently and it was just down to the observer of an experiment to see the underlying truth. Quantum mechanics showed that at an atomic level, such a deterministic view was meaningless and one could only talk about probabilities of outcomes instead. We could no longer talk about cause and effect but only chance. Einstein, and many others, found this hard to accept, but had to agree that this is what the equations showed. For the first time, physics moved well beyond the laboratory of experience and firmly into the realm of abstract mathematics.

the condensed idea
Know your limits

27 Copenhagen interpretation

The equations of quantum mechanics gave scientists the right answers, but what did they mean? Danish physicist Niels Bohr developed the Copenhagen interpretation of quantum mechanics, blending the wave equation of Schrödinger and the uncertainty principle of Heisenberg. Bohr argued that there is no such thing as an isolated experiment – that the observer's interventions fix the outcomes of quantum experiments. In doing so, he challenged the very objectivity of science.

In 1927, competing views of quantum mechanics were rife. Erwin Schrödinger argued that the physics of waves underlay quantum behaviour, which could all be described using wave equations. Werner Heisenberg, on the other hand, believed that the particle nature of electromagnetic waves and matter, described in his tabular matrix representation, was of the foremost importance in comprehending nature. Heisenberg had also shown that our understanding was fundamentally limited by his uncertainty principle. He believed that both the past and future were unknowable until fixed by observation because of the intrinsic uncertainty of all the parameters describing a subatomic particle's movement.

Another man tried to pull together all the experiments and theories to form a new picture that could explain the whole. This was Niels Bohr, the head of Heisenberg's department at the University of Copenhagen, and the scientist who had explained the quantum energy states of electrons in the hydrogen atom. Bohr, together with Heisenberg, Max Born and others,

timeline

AD 1901
Planck publishes his law of black-body radiation

1905
Einstein uses light quanta to explain the photoelectric effect

NIELS BOHR 1885–1962

Niels Bohr lived through two world wars, and worked with some of the best physicists around. Young Niels pursued physics at Copenhagen University, performing award-winning physics experiments in his father's physiology laboratory. He moved to England after his doctorate but clashed with J.J. Thomson. After working with Ernest Rutherford in Manchester, he returned to Copenhagen, completing his work on the 'Bohr atom' (still how most people picture an atom today). He won the Nobel Prize in 1922, just before quantum mechanics fully appeared. To escape Hitler's Germany in the 1930s, scientists flocked to Bohr's Institute of Theoretical Physics in Copenhagen, where they were entertained in a mansion donated by Carlsberg, the Danish brewers. When the Nazis occupied Denmark in 1940, Bohr fled via a fishing boat to Sweden and then England.

developed a holistic view of quantum mechanics that became known as the Copenhagen interpretation. It is still the favourite interpretation of most physicists, although other variations have been suggested.

Two sides Niels Bohr brought a philosophical approach to bear on the new science. In particular, he highlighted the impact that the observer themselves has on the outcomes of quantum experiments. First he accepted the idea of 'complementarity', that the wave and particle sides of matter and light were two faces of the same underlying phenomenon and not two separate families of events. Just as pictures in a psychological test can switch appearance depending on how you look at them – two wiggly mirrored lines appearing either as a vase outline or two faces looking at one another – wave and particle properties were complementary ways of seeing the same phenomenon. It was not light that changed its character, but rather how we decided to view it.

To bridge the gap between quantum and normal systems, including our own experiences on human scales, Bohr also introduced the 'correspondence principle', that quantum behaviour must disappear for

1927

Heisenberg publishes his uncertainty principle

Copenhagen interpretation formulated

larger systems that we are familiar with, when Newtonian physics is adequate.

Unknowability Bohr realized the central importance of the uncertainty principle, which states that one cannot measure both the position and momentum (or speed) of any subatomic particle at the same time. If one quantity is measured accurately, then the other is inherently uncertain. Heisenberg thought that the uncertainty came about because of the mechanics of the measurement act itself. To measure some thing, even to look at it, we must bounce photons of light off it. Because this always involves the transfer of some momentum or energy, then this act of observation disturbed the original particle's motion.

> ‘We are in a jungle and find our way by trial and error, building our road behind us as we proceed.’
>
> **Max Born, 1882–1970**

Bohr, on the other hand, thought Heisenberg's explanation flawed. He argued that we can never completely separate the observer from the system he or she is measuring. It was the act of observation itself that set the system's final behaviour, through the probabilistic wave–particle behaviour of quantum physics and not due to simple energy transfer. Bohr thought that an entire system's behaviour needed to be considered as one; you could not separate the particle, the radar and even the observer themselves. Even if we look at an apple, we need to consider the quantum properties of the whole system, including the visual system in our own brain that processes the photons from the apple.

Bohr also argued that the very word ‘observer’ is wrong because it conjures up a picture of an external viewer separated from the world that is being watched. A photographer such as Ansel Adams may capture the pristine natural beauty of the Yosemite wilderness, but is it really untouched by man? How can it be if the photographer himself is there too? The real picture is of a man standing within nature, not separate from it. To Bohr, the observer was very much part of the experiment.

This concept of observer participation was shocking to physicists, because it challenged the very way that their science had always been done and the fundamental concept of scientific objectivity. Philosophers also balked. Nature was no longer mechanical and predictable but, deep down, was inherently unknowable. What did this mean for concepts of basic truth, let alone simple ideas such as past and future? Einstein, Schrödinger and others had difficulty dropping their firm beliefs in an external, deterministic and verifiable universe. Einstein believed that, because it could only be described with statistics, the theory of quantum mechanics must be at the every least incomplete.

Collapsing wave functions Given that we observe subatomic particles and waves as either one or the other entity, what decides how they manifest themselves? Why does light passing through two slits interfere like waves on Monday, but switch to particle-like behaviour on Tuesday if we try to catch the photon as it passes through one slit? According to Bohr and supporters of the Copenhagen interpretation, the light exists in both states simultaneously, both as a wave and a particle. It only dresses itself as one or the other when it is measured. So we choose in advance how it turns out by deciding how we would like to measure it.

At this point of decision-making, when the particle- or wave-like character is fixed, we say that the wave function has collapsed. All the probabilities for outcomes that are contained in Schrödinger's wave function description crush down so that everything apart from the eventual outcome is lost. So, according to Bohr, the original wave function for a beam of light contains all the possibilities within it, whether the light appears in its wave or particle guise. When we measure it, it appears in one form, not because it changes from one type of substance to another but because it is truly both at the same time. Quantum apples and oranges are neither, but instead a hybrid.

Physicists still have trouble in understanding intuitively what quantum mechanics means and others since Bohr have offered new ways of interpreting it. Bohr argued that we needed to go back to the drawing board to understand the quantum world, and could not use concepts that were familiar in everyday life. The quantum world is something else strange and unfamiliar, and we must accept that.

‘Anyone who is not shocked by quantum theory has not understood it.’
Niels Bohr, 1885–1962

the condensed idea
You choose

28 Schrödinger's cat

Schrödinger's cat is both alive and dead at the same time. In this hypothetical experiment, a cat sitting inside a box may or may not have been killed by a poison capsule, depending on some random trigger. Erwin Schrödinger used this metaphor to show how ridiculous he found the Copenhagen interpretation of quantum theory, which predicted that, until the outcome was actually observed, the cat should be in a state of limbo, both alive and dead.

In the Copenhagen interpretation of quantum theory, quantum systems exist as a cloud of probability until an observer flicks the switch and selects one outcome for his or her experiment. Before being observed, the system takes on all possibilities. Light is both particle and wave until we decide which form we want to measure – then it adopts that form.

While a probability cloud may sound like a plausible concept for an abstract quantity like a photon or light wave, what might it mean for something larger that we might be able to be aware of? What really is the nature of this quantum fuzziness?

In 1935, Erwin Schrödinger published an article containing a hypothetical experiment which tried to illustrate this behaviour with a more colourful and familiar example than subatomic particles. Schrödinger was highly critical of the Copenhagen view that the act of observation influenced its behaviour. He wanted to show how daft the Copenhagen interpretation was.

AD 1927
The Copenhagen interpretation of quantum mechanics

1935
Schrödinger's quantum cat experiment is proposed

Quantum limbo Schrödinger considered the following situation, which was entirely imaginary. No animals were harmed.

'*A cat is penned up in a steel chamber, along with the following diabolical device (which must be secured against direct interference by the cat): in a Geiger counter there is a tiny bit of radioactive substance, so small that perhaps in the course of one hour one of the atoms decays, but also, with equal probability, perhaps none; if it happens, the counter tube discharges and through a relay releases a hammer which shatters a small flask of hydrocyanic acid. If one has left this entire system to itself for an hour, one would say that the cat still lives if meanwhile no atom has decayed. The first atomic decay would have poisoned it.*'

So there is a 50:50 probability of the cat being either alive (hopefully) or dead when the box is opened after that time. Schrödinger argued that, following the logic of the Copenhagen interpretation, we would have to think of the cat as existing in a fuzzy blend of states, being both alive and dead at the same time, while the box was closed. Just as the wave or particle view of an electron is only fixed on the point of detection, the cat's future is only determined when we choose to open the box and view it. On opening the box we make the observation and the outcome is set.

Surely, Schrödinger grumbled, this was ridiculous and especially so for a real animal such as a cat. From our everyday experience we know that the cat must be either alive or dead, not a mixture of both, and it is madness to imagine that it was in some limbo state just because we were not looking at it. If the cat lived, all it would remember was sitting in the box being very much alive, not being a probability cloud or wave function.

Amongst others, Einstein agreed with Schrödinger that the Copenhagen picture was absurd. Together they posed further questions. As an animal, was the cat able to observe itself, and so collapse its own wave function? What does it take to be an observer? Need the observer be a conscious being like a human or would any animal do? How about a bacterium?

1957

Everett suggests the many worlds hypothesis

Going even further, we might question whether anything in the world exists independently of our observation of it. If we ignore the cat in the box and just think of the decaying radioactive particle, will it have decayed or not if we keep the box closed? Or is it in quantum limbo until we open the box's flap, as the Copenhagen interpretation requires? Perhaps the entire world is in a mixed fuzzy state and that nothing resolves itself until we observe it, causing the wave function to collapse when we do. Does your workplace disintegrate when you are away from it at weekends, or is it protected by the gazes of passers by? If no one is watching, does your holiday home in the woods cease to exist in reality? Or does it wait in a blend of probability states, as a superposition of having being burned down, flooded, invaded by ants or bears, or sitting there just fine, until you return to it? Do the birds and squirrels count as observers? As odd as it is, this is how Bohr's Copenhagen interpretation explains the world on the atomic scale.

Many worlds The philosophical problem of how observations resolve outcomes has led to another variation on the interpretation of quantum

ERWIN SCHRÖDINGER 1887–1961

Austrian physicist Erwin Schrödinger pursued quantum mechanics and tried (and failed), with Einstein, to unify gravity and quantum mechanics into a single theory. He favoured wave interpretations and disliked wave–particle duality, leading him into conflict with other physicists.

As a boy Schrödinger loved German poetry but nevertheless decided to pursue theoretical physics at university. Serving on the Italian front during the First World War, Schrödinger continued his work remotely and even published papers, returning afterwards to academia. Schrödinger proposed his wave equation in 1926, for which he was awarded the Nobel Prize with Paul Dirac in 1933. Schrödinger then moved to Berlin to head Max Planck's old department, but with Hitler's coming to power in 1933 he decided to leave Germany. He found it hard to settle, and worked for periods in Oxford, Princeton and Graz. With Austria's annexation in 1938, he fled again, moving finally to a bespoke position created for him at the new Institute for Advanced Studies in Dublin, Ireland, where he remained until retiring to Vienna. Schrödinger's personal life was as complicated as his professional life; he fathered children with several women, one of whom lived with him and his wife for a time in Oxford.

theory – the many worlds hypothesis. Suggested in 1957 by Hugh Everett, the alternative view avoids the indeterminacy of unobserved wave functions by saying instead that there are an infinite number of parallel universes. Every time an observation is made, and a specific outcome noted, a new universe splits off. Each universe is exactly the same as the other, apart from the one thing that has been seen to change. So the probabilities are all the same, but the occurrence of events moves us on through a series of branching universes.

In a many worlds interpretation of Schrödinger's cat experiment, when the box is opened the cat is no longer in a superposition of all possible states. Instead it is either alive in one universe or dead in another parallel one. In one universe the poison is released, in the other it is not.

Whether this is an improvement on being in wave function limbo is arguable. We may well avoid the need for an observer to pull us out of being just a probability cloud sometimes, but the cost is to invoke a whole range of alternative universes where things are very slightly different. In one universe I am a rock star, in another just a busker. Or in one I am wearing black socks, in another grey. This seems a waste of a lot of good universes (and hints at universes where people have garish wardrobes). Other alternative universes might be more significant – in one Elvis still lives, in another John F. Kennedy wasn't shot, in another Al Gore was President of the USA. This idea has been borrowed widely as a plot device in movies, such as *Sliding Doors* where Gwyneth Paltrow lives two parallel lives in London, one successful, one not.

Some physicists today argue that Schrödinger's thinking on his metaphorical cat experiment was invalid. Just as with his exclusively wave-based theory, he was trying to apply familiar physics ideas to the weird quantum world, when we just have to accept that it is strange down there.

the condensed idea
Dead or alive?

29 The EPR paradox

Quantum mechanics suggests information can be transmitted instantaneously between systems, no matter how far apart they are. Such entanglement implies vast webs of interconnectivity between particles across the universe. Einstein, Podolsky and Rosen thought this absurd and queried this interpretation in their paradox. Experiments show quantum entanglement is true, opening up applications in quantum cryptography, computing and even teleportation.

Albert Einstein never accepted the Copenhagen interpretation of quantum mechanics, which claimed quantum systems existed in probabilistic limbo until observed, when they adopt their final state. Before being filtered by observation, the system exists in a combination of all possible states of being. Einstein was unhappy with this picture, arguing such a blend was unrealistic.

Paradoxical particles In 1935, Einstein together with Boris Podolsky and Nathan Rosen encapsulated their discomfort in the form of a paradox. This became known as the Einstein–Podolsky–Rosen, or EPR, paradox. Imagine a particle that decays into two smaller ones. If the original mother particle was stationary, the daughter particles must have equal and opposite linear and angular momentum, so that the sum is zero (as these are conserved). So the emerging particles must fly apart and spin in opposite directions. Other quantum properties of the pair are similarly

> ‘I, at any rate, am convinced that He [God] does not throw dice.’
>
> **Albert Einstein, 1926**

timeline

AD 1927
The Copenhagen interpretation is proposed

1935
Einstein, Podolsky and Rosen set out their paradox

Teleportation

Teleportation is widely depicted in science fiction. The beginnings of communication technologies, such as the telegraph in the 19th century, raised the prospect that information other than electrical pulses could be transferred over vast distances. In the 1920s and 1930s, teleportation began to appear in books, such as those by Arthur Conan Doyle, and became a staple of science fiction stories. In George Langelaan's *The Fly* (adapted into three film versions), a scientist teleports himself but his body's information becomes mixed up with that of a housefly, turning him into a part human, part fly chimera. Teleportation really took off with the cult TV show *Star Trek*, which included the famous line 'beam me up Scottie'. The starship *Enterprise* teleporter dismantles the transmittee atom by atom and reassembles them perfectly. In real life, teleportation was thought impossible due to Heisenberg's uncertainty principle. Although it is impossible to transmit actual atoms, quantum entanglement enables the long distance transmission of information, but so far this has only worked for tiny particles.

linked. Once emitted, if we were to measure the spin direction of one of the particles, we would immediately also know that the other member of the pair has the opposite spin – even if a significant time had elapsed and it was far away or out of reach. It is like looking at an identical twin and noticing her eye colour. If it is green then, at that instant, we know the other twin has green eyes too.

Explaining this using the Copenhagen interpretation, you would say that, before any measurement, both particles (or twins) existed in a superposition of both possible states. The particles' wave functions included information about them spinning in either direction; the twins had a mixture of all possible eye colours. When we measure one member of the pair, wave functions for both must collapse at the same time. Einstein, Podolsky and Rosen argued that this made no sense. How could

1964
John Bell derives inequalities for a local reality

1981–2
Bell's inequalities are shown to be violated, supporting entanglement

1993
Quantum bits are christened qubits

you affect a particle instantaneously that could potentially be far away from its companion? Einstein had already shown that the speed of light was the universal speed limit; nothing could travel faster. How was the fact of the first particle's observation communicated to the second? That a measurement on one side of the universe could 'simultaneously' affect matter on the opposite side must mean quantum mechanics was wrong.

Entanglement In the same paper in which he described his cat paradox, Schrödinger used 'entanglement' to describe this weird action at a distance.

To Bohr, it was inevitable that the universe was linked together at the quantum level. But Einstein preferred to believe in a 'local reality' where knowledge about the world locally was certain. Just as the twins were presumably born with the same eye colour, and weren't walking around in a fuzzy multicoloured-eye state until we observed them, Einstein presumed that the particle pair was emitted in one or other way which was fixed thereafter; there need be no communication at a distance or role for the observer. Einstein supposed some hidden variables, now reformulated as 'Bell's inequalities', would be found that would eventually prove him correct, but no evidence has been found to support this idea.

Einstein's idea of local reality has been shown to be false. Experiments have even demontrated quantum entanglement is true, even where there are more than two particles and for entangled particles separated by many kilometres.

Quantum information Quantum entanglement originally began as a philosophical debate but now allows the encoding and transmittal of information in ways that are unlike anything possible before. In normal computers, information is encoded as bits with fixed values in binary code. In quantum encoding, two or more quantum states are used, but the system can also exist in a blend of these states. In 1993, the term 'qubit' was coined as shorthand for a quantum bit (the quantum blends of bit values) and quantum computers are now being designed on these principles.

Entangled states provide novel communication links between the qubits. If a measurement occurs, then this begins a cascade of further quantum communications between elements of the system. Measurement of one element sets the values of all the others; such effects are useful in quantum cryptography and even quantum teleportation.

The indeterminacy of quantum mechanics actually rules out teleportation as depicted in much science fiction, whereby a scientist takes all the information from something and reassembles it elsewhere. We can't obtain all the information because of the uncertainty principle. So teleporting a human, or even a fly, is impossible. A quantum version however is possible by manipulating entangled systems. If two people, often named Alice and Bob by physicists, share a pair of entangled photons, Alice can make measurements of her photon so as to transfer all the original information to Bob's entangled photon. Bob's photon becomes indistinguishable from her original, although it is a reproduction. Whether or not this is true teleportation is a good question. No photons or information travelled anywhere, so Alice and Bob could be on opposite sides of the universe and still transform their entangled photons.

'It seems that even God is bound by the uncertainty principle, and can not know both the position, and the speed, of a particle. So does God play dice with the universe? All the evidence points to him being an inveterate gambler, who throws the dice on every possible occasion.'

Stephen Hawking, 1993

Quantum cryptography is based on the use of quantum entanglement as a linked encryption key. The sender and receiver must each hold the components of an entangled system. A message can be scrambled at random, and the unique code to unravel it sent via quantum entanglement connections to the receiver. This has the advantage that if the message is intercepted, any measurement ruins the message (changing its quantum state), so it can be used once only and only read by someone who knows exactly what quantum measurements should be performed to reveal it via the key.

Entanglement tells us that it is simply wrong to assume that our entire world exists independently in one form, irrespective of our measurement of it. There is no such thing as an object fixed in space, just information. We can only gather information about our world and order it as we see fit, so that it makes sense to us. The universe is a sea of information; the form we assign to it is secondary.

the condensed idea
Instant messaging

30 Pauli's exclusion principle

Pauli's exclusion principle explains why matter is rigid and impermeable – why we do not sink through the floor or push our hand through a table. It's also responsible for the existence of neutron stars and white dwarfs. Wolfgang Pauli's rules apply to electrons, protons and neutrons, so affecting all matter. The principle states that none of these particles can have the same set of quantum numbers simultaneously.

What gives matter its rigidity? Atoms are mostly empty space, so why can't you squeeze them like a sponge or push materials through each other like cheese through a grater? The question of why matter inhabits space is one of the most profound in physics. If it was not true we could fall into the centre of the earth or sink through floors, and buildings would squash under their own weight.

Not the same The Pauli exclusion principle, devised by Wolfgang Pauli in 1925, explains why normal atoms cannot coexist in the same region of space. Pauli suggested that the quantum behaviour of atoms and particles meant that they had to follow certain rules forbidding them from having the same wave function or, equivalently, the same quantum properties. Pauli devised his principle to try to explain the behaviour of electrons in atoms. Electrons were known to prefer certain energy states, or

timeline

AD 1925
Pauli proposes his exclusion principle

1933
The neutron is discovered and neutron stars predicted

shells, around the nucleus. But the electrons were spread between these shells and never all congregated in the lowest energy shell. They seemed to populate the shells according to rules that Pauli worked out.

Just as Newton's physics is expressed in terms of force, momentum and energy, quantum mechanics has its own set of parameters. Quantum spin is analogous to angular momentum, for example, but is quantized and takes on only certain values. By solving Schrödinger's equation, four quantum numbers are needed to describe any particle – three spatial coordinates and the fourth one, spin. Pauli's rules stated that no two electrons in an atom could have the same four quantum numbers. No two electrons could be in the same place with the same properties at the same time. So as the number of electrons in an atom grows, as the atoms become heavier for instance, the electrons fill up their allotted spaces and gradually move out to higher and higher shells. It is like seats in a tiny theatre filling up, from the stage outwards.

Fermions Pauli's rules apply to all electrons and other particles whose quantum spin comes in half-integer multiples of the basic unit, including the proton and neutron. Such particles are called 'fermions' after Italian physicist Enrico Fermi. Fermions have asymmetric wave functions, switching from positive to negative, as expressed in Schrödinger's equation. Spin also has a direction, so fermions can lie next to one another if they possess opposite spin. Two electrons may both inhabit the lowest energy state of an atom but only if their spins are misaligned.

Because the basic building blocks of matter, electrons, protons and neutrons are all fermions, Pauli's exclusion principle dictates the behaviour of atoms. As none of these particles can share its quantum energy state with any other, atoms are inherently rigid. Electrons distributed across many energy shells cannot all be squashed down into the shell closest to the nucleus; in fact, they resist this compression with great pressure. So, no two fermions can sit in the same theatre seat.

1967

The first pulsar, a type of neutron star, is discovered

WOLFGANG PAULI 1900–59

Wolfgang Pauli is best known for his exclusion principle and for proposing the existence of the neutrino. Pauli was a precocious student in Austria, reading Einstein's work and writing papers on relativity. Heisenberg described Pauli as a night bird who worked in cafes and rarely attended morning lectures. Pauli suffered many personal problems, including the suicide of his mother, a short failed marriage and a drink problem. For help, he consulted Swiss psychologist Carl Jung, who recorded thousands of Pauli's dreams. Pauli's life picked up when he remarried, but then came the Second World War. From the United States he worked to keep European science alive. He returned to Zurich after the war, and was awarded the Nobel Prize in 1945. In later years, he pursued the more philosophical aspects of quantum mechanics and its parallels in psychology.

Quantum crush Neutron stars and white dwarfs owe their very existence to the Pauli exclusion principle. When a star reaches the end of its life, and can no longer burn fuel, it implodes. Its own enormous gravity pulls all the gas layers inwards. As it collapses, some of the gas can be blasted away (as in a supernova explosion), but the remaining embers contract even more. As the atoms are crushed together, the electrons try to resist compaction. They inhabit the innermost energy shells they can without violating Pauli's principle, propping up the star due to this 'degeneracy pressure' alone. White dwarfs are stars of about the Sun's mass, squashed down into a region of similar radius to the Earth. They are so dense that one sugar cube worth of white dwarf material can weigh a tonne.

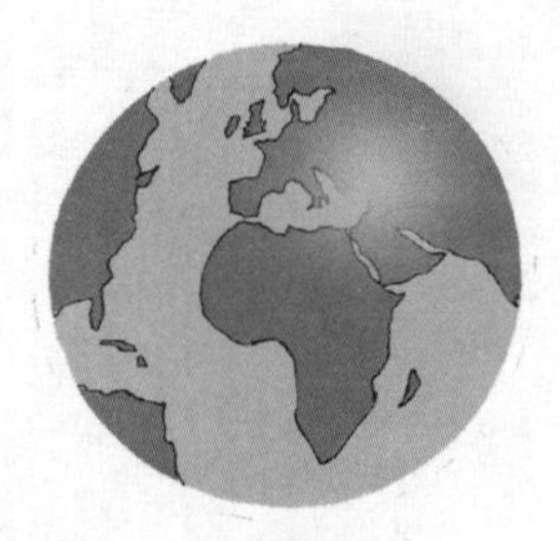

Earth

White dwarf

Neutron star

For stars with larger self-gravity, in particular for stars more than 1.4 times the Sun's mass (called the Chandrasekhar mass limit), the compaction does not stop there. In a second process, the protons and electrons can merge to form neutrons, so the giant star reduces itself to a tight ball of neutrons.

As before, because the neutrons are fermions they cannot all have the same quantum state. Degeneracy pressure again holds up the star, but this time it is confined to a radius of only ten or so kilometres, squashing the entire mass of the Sun, or several suns, into a region the length of

> **"The question, as to why all electrons for an atom in its ground state were not bound in the innermost shell, had already been emphasized by Bohr as a fundamental problem . . . no explanation of this phenomenon could be given on the basis of classical mechanics."**
>
> **Wolfgang Pauli, 1945**

Manhattan. Neutron stars are so dense that a sugar-cube-sized block would weigh more than a hundred million tonnes. In the event that the gravity exceeds even this, such as for the largest stars, further compaction ultimately produces a black hole.

Bosons Pauli's rules apply only to fermions. Particles with integer multiples of the basic unit of spin and symmetric wave functions are called 'bosons' after Indian physicist Satyendranath Bose. Bosons include particles associated with the fundamental forces, such as photons, and some symmetric nuclei such as that of helium (which contains two protons and two neutrons). Any number of bosons can occupy the same quantum state and this can lead to coordinated group behaviour. One example is the laser, whereby many photons of a single colour all act together.

Originally an extension to Bohr's picture of the atom, Pauli's exclusion principle just preceded the main advance of quantum theory championed by Heisenberg and Schrödinger. But it is fundamental to the study of the atomic world and, unlike much of quantum mechanics, has consequences that we can actually touch.

the condensed idea
Is this seat taken?

31 Superconductivity

At very low temperatures, some metals and alloys conduct electricity without any resistance. The current in these superconductors can flow for billions of years without losing any energy. As electrons become coupled and all move together, avoiding the collisions that cause electrical resistance, they approach a state of perpetual motion.

When mercury is cooled to a few degrees above absolute zero, it conducts electricity with no resistance whatsoever. This was discovered in 1911 by Dutch physicist Heike Onnes, when he dropped mercury into liquid helium at a temperature of 4.2 K (degrees above absolute zero). With no resistance to current, the first superconducting material was found. Soon afterwards, similar behaviour was seen for other cold metals including lead and compounds such as niobium nitride. All resistance disappeared below some critical temperature that varied for different materials.

Perpetual motion A consequence of zero resistance is that a current set flowing through a superconductor can flow forever. In the laboratory, currents have been maintained for many years, and physicists estimate such a current would last for billions of years before losing any energy. It is as close to perpetual motion that scientists have come.

Group think Physicists puzzled over how such a major transition could happen at cool temperatures. The critical temperature suggested a quick phase transition, so physicists looked at the quantum behaviour of electrons in a metal. Quantum mechanics gave some clues, and various ideas were put forward in the 1950s. In 1957, American physicists John Bardeen, Leon Cooper and John Schrieffer came up with a convincing and

timeline

AD 1911 Onnes discovers superconductivity

1925 Bose–Einstein condensates are predicted

1933 Superconductors are shown to repel magnetic fields

1940s Superconducting compounds are discovered

Superfluids

Superfluids are fluids that have no viscosity so they can flow through a tube forever without any friction. Superfluidity has been known since the 1930s. One example is super-cooled helium-4 (atomic weight 4, made of two protons and two neutrons and two electrons). Helium-4 atoms are bosons, made of pairs of fermions.

Superfluids behave very strangely when placed in a container – they can flow, in a layer one atom thick, up the side of the container. A fountain can be created if a capillary tube is inserted and warmed, because the superfluid cannot hold a temperature gradient (it has infinite thermal conductivity) and the heat immediately causes a pressure change. If you tried to spin a bucket of superfluid (see page 4), something weird happens. Since there is no viscosity, the fluid does not immediately rotate but remains still. If you spin the bucket faster, then at some critical point the superfluid suddenly starts to spin. Its speed is quantized – the superfluid only rotates at certain values.

complete explanation of superconductivity in metals and simple alloys, now called BCS theory. It suggested that superconductivity happens because of the weird behaviour of electrons when they are linked in pairs.

The electron pairs, called Cooper pairs, interact with the lattice of metal atoms via vibrations that tie them together. A metal is a lattice of positive charged nuclei about which a 'sea' of electrons is free to float. When the metal is very cold, and the lattice is still, a passing negatively charged electron tugs a little on the positive points of the lattice and pulls them out into a wake-like ripple. Another electron moving nearby can become attracted to this region of slightly more intense positive charge, and the two electrons become coupled. The second one follows the other around. This happens to electrons right across the metal, and many synchronized electron pairs link together into a moving wave pattern.

1957
The BCS theory of superconductivity is proposed

1986
High-temperature superconductors are created

1995
Bose–Einstein condensates are made in the laboratory

Bose-Einstein condensates

At ultracold temperatures, groups of bosons can behave very strangely. Near absolute zero, many bosons can all inhabit the same quantum state, making quantum behaviour visible on much larger scales. First predicted by Albert Einstein in 1925, and based on ideas by Indian physicist Satyendranath Bose, these so-called Bose–Einstein condensates (BECs) were not created in the laboratory until 1995. Eric Cornell and Carl Wieman of the University of Colorado, and a little later Wolfgang Ketterle of MIT, saw this behaviour in a gas of rubidium atoms cooled to 170 billionths of a kelvin. In the BEC, all the clustered atoms have the same velocity, blurred only by the Heisenberg uncertainty principle. BECs behave as superfluids. Bosons are allowed to share quantum states with each other. Einstein speculated that cooling bosons to below a very low critical temperature would cause them to fall (or 'condense') into the lowest-energy quantum state, resulting in a new form of matter. BECs disrupt very easily, so it is still early for practical applications, but they can teach us much about quantum mechanics.

A single electron must follow Pauli's exclusion principle, that forbids such particles with asymmetric wave functions (fermions) from sharing the same quantum state. Where there are many electrons, therefore, if they are in the same region they must have different energies from one another. This is what normally happens in an atom or metal. But when electrons are paired together and behave as a single particle, they no longer follow this behaviour. Their overall wave function becomes symmetric and together they are no longer fermions, but rather bosons. And as bosons, the electron pairs can share the same minimum energy. This results in the sets of pairs having slightly lower overall energy than free electrons would in the metal. It is this particular energy difference that creates the quick transition in properties at the critical temperature.

When the heat energy of the lattice is less than this energy drop, we see the steady flow of electron pairs coupled to the lattice vibrations that characterizes superconductivity. Because the lattice waves drive the motions over large distances through the lattice, there is no resistance – all the electron pairs are moving with respect to one another. Avoiding any collisions with the still lattice atoms, the electron pairs act as a super fluid that can flow unimpeded. At warmer temperatures, the Cooper pairs break up and lose their boson-like properties. The electrons can collide with the lattice ions, which are now warm and vibrating, and create electrical resistance. The quick transition switches between the states when electrons change from coordinated flows of bosons into erratic fermions or vice versa.

Warm superconductors In the 1980s, superconductor technology took off. In 1986, Swiss researchers discovered a new class of ceramic materials that became superconductors at relatively warm temperatures – so-called 'high-temperature superconductors'. Their first compound, a combination of lanthanum, barium, copper and oxygen (known as copper oxides or cuprates), transitioned to superconducting behaviour at 30 kelvins. A year later, others designed a material that became a superconductor at temperatures of about 90 kelvins, warmer than the widely used coolant liquid nitrogen. Using perovskite-based ceramics and (thallium-doped) mercuric-cuprates, superconducting temperatures have now reached around 140 kelvins and even higher critical tempratures are attainable at high pressure.

Such ceramics were supposed to be insulators, so this was unexpected. Physicists are still searching for a new theory to explain high-temperature superconductivity. Nevertheless, developing them is now a fast evolving field of physics that would revolutionize electronics.

What are superconductors used for? They help make powerful electromagnets, as used in MRI scanners in hospitals and particle accelerators. One day they could be used for efficient transformers or even for magnetic levitating trains. But because they currently work at ultra low temperatures, their uses are somewhat limited. Hence the search is on for high-temperature superconductors that might have dramatic implementations.

the condensed idea
Resistance is futile

32 Rutherford's atom

Atoms are not the smallest building blocks of matter once thought. Early in the 20th century, physicists such as Ernest Rutherford broke into them, revealing first layers of electrons and then a hard core, or nucleus, of protons and neutrons. To bind the nucleus together a new fundamental force – the strong nuclear force – was invented. The atomic age had begun.

The idea that matter is made up of swarms of tiny atoms has been around since the Greeks. But whereas the Greeks thought the atom was the smallest indivisible component of matter, 20th-century physicists realized this was not so and began to probe the inner structure of the atom itself.

Plum pudding model The first layer to be tackled was that of the electron. Electrons were liberated from atoms in 1887 by Joseph John Thomson who fired an electric current through gas contained in a glass tube. In 1904, Thomson proposed the 'plum pudding model' of the atom, where negatively charged electrons were sprinkled like prunes or raisins through a sponge dough of positive charge. Today it might have been called the blueberry muffin model. Thomson's atom was essentially a cloud of positive charge containing electrons, which could be set free relatively easily. Both the electrons and positive charges could mix throughout the 'pudding'.

The nucleus Not long after, in 1909, Ernest Rutherford puzzled over the outcome of an experiment he had performed in which heavy alpha particles were fired through very thin gold foil, thin enough that most of

timeline

AD **1887**
Thomson discovers the electron

1904
Thomson proposes the plum pudding model

1909
Rutherford performs his gold foil experiment

ERNEST RUTHERFORD 1871–1937

New Zealander Rutherford was a modern-day alchemist, transmuting one element, nitrogen, into another, oxygen, through radioactivity. An inspiring leader of the Cavendish Laboratory in Cambridge, England, he mentored numerous future Nobel Prize winners. He was nicknamed 'the crocodile', and this animal is the symbol of the laboratory even today. In 1910, his investigations into the scattering of alpha rays and the nature of the inner structure of the atom led him to identify the nucleus.

the particles passed straight through. To Rutherford's astonishment a small fraction of particles ricocheted straight back off the foil, heading towards him. They reversed direction by 180 degrees, as if they had hit a brick wall. He realized that within the gold atoms that made up the foil sheet lay something hard and massive enough to repel the heavy alpha particles.

Rutherford understood that Thomson's plum pudding model could not explain this. If an atom was just a paste of mixed positively and negatively charged particles then none would be heavy enough to knock back the bigger alpha particle. So, he reasoned, the gold atoms must have a dense core, called the 'nucleus' after the Latin word for the 'kernel' of a nut. Here began the field of nuclear physics, the physics of the atomic nucleus.

> **'It was almost as incredible as if you fired a 15-inch shell at a piece of tissue paper and it came back to hit you.'**
>
> **Ernest Rutherford, 1936**

Isotopes Physicists knew how to work out the masses of different elements of the periodic table, so they knew the relative weights of atoms. But it was harder to see how the charges were arranged. Because Rutherford only knew about electrons and the positively charged nucleus, he tried to balance the charges by assuming that the nucleus was made up of a mix of protons (positively charged particles that he discovered in 1918

1911
Rutherford proposes the nuclear model

1918
Rutherford isolates the proton

1932
Chadwick discovers the neutron

1934
Yukawa proposes the strong nuclear force

Three of a Kind

Radioactive substances emit three types of radiation, called alpha, beta and gamma radiation. Alpha radiation consists of heavy helium nuclei comprising two protons and two neutrons bound together. Because they are heavy, alpha particles do not travel far before losing their energy in collisions and can be stopped easily, even by a piece of paper. A second type of radiation is carried by beta particles; these are high-speed electrons – very light and negatively charged. Beta particles can travel further than alpha radiation but may be halted by metal such as an aluminium plate. Third are gamma rays, which are electromagnetic waves, associated with photons, and so carry no mass but a lot of energy. Gamma rays are pervasive and can be shielded only with dense blocks of concrete or lead. All three types of radiation are emitted by unstable atoms that we describe as radioactive.

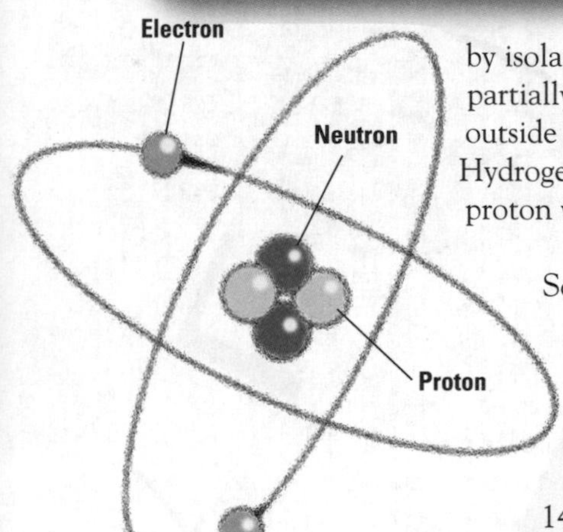

by isolating the nuclei of hydrogen) and some electrons that partially neutralized the charge. The remaining electrons circled outside the nucleus in the familiar orbitals of quantum theory. Hydrogen, the lightest element, has a nucleus containing just one proton with one electron orbiting it.

Some other forms of elements with odd weights were known, called isotopes. Carbon usually has a weight of 12 atomic units, but is occasionally seen with a weight of 14 units. Carbon-14 is unstable with a half-life (the time it takes for half the atoms to decay by emitting a radioactive particle) of 5730 years, emitting a beta particle to become nitrogen-14. This reaction is used in radiocarbon dating to measure the ages of archaeological artefacts thousands of years old, such as wood or charcoal from fires.

Neutrons In the early 1930s a new type of 'radiation' was found that was heavy enough to free protons from paraffin but with no charge. Cambridge physicist James Chadwick showed that this new radiation was in fact a neutral particle with the same mass as the proton. It was named the neutron and the model of the atom was rearranged. Scientists realized that a carbon-12 atom, for instance, contains 6 protons and 6 neutrons in the nucleus (to give it a mass of 12 atomic units) and 6 orbiting electrons. Neutrons and protons are known as nucleons.

> **'Nothing exists except atoms and empty space; everything else is opinion.'**
>
> **Democritus, 460–370 BC**

Strong force The nucleus is absolutely tiny compared with the full extent of the atom and its veiling electrons. A hundred thousand times smaller than an atom, the nucleus is only a few femtometres (10^{-15} metres, or one ten million billionth of a metre) across. If the atom were scaled up to the diameter of the Earth, the nucleus at the centre would be just 10 kilometres wide, or the length of Manhattan. The nucleus harbours practically all the mass of the atom in one tiny spot, and this can include many tens of protons. What holds all this positive charge together in such a small space so tightly? To overcome the electrostatic repulsion of the positive charges and bind the nucleus together, physicists had to invent a new type of force, called the strong nuclear force.

If two protons are brought up close to one another, they initially repel because of their like charges (following Maxwell's inverse square law). But if they are pushed even closer the strong nuclear force locks them together. The strong force only appears at very small separations, but it is much greater than the electrostatic force. If the protons are pushed even closer together, they resist, acting as if they are hard spheres – so there is a firm limit to how close together they can go. This behaviour means that the nucleus is tightly bound, very compact and rock hard.

In 1934, Hideki Yukawa proposed that the nuclear force was carried by special particles (called mesons), that act in a similar way to photons. Protons and neutrons are glued together by exchanging mesons. Even now it is a mystery why the strong nuclear force acts on such a specific distance scale – why it is so weak outside the nucleus and so strong at close range. It is as if it locks the nucleons together at a precise distance. The strong nuclear force is one of the four fundamental forces, along with gravity, electromagnetism and another nuclear force called the weak force.

the condensed idea
The hard core

33 Antimatter

Fictional spaceships are often powered by 'antimatter drives', yet antimatter itself is real and has even been made artificially on Earth. A mirror image form of matter that has negative energy, antimatter cannot coexist with matter for long – both annihilate in a flash of energy if they come into contact. The very existence of antimatter hints at deep symmetries in particle physics.

Walking down the street you meet a replica of yourself. It is your antimatter twin. Do you shake hands? Antimatter was predicted in the 1920s and discovered in the 1930s by bringing together quantum theory and relativity. It is a mirror image form of matter, where particles' charges, energies and other quantum properties are all reversed in sign. So an anti-electron, called a positron, has the same mass as the electron but instead has a positive charge. Similarly, protons and other particles have opposite antimatter siblings.

> **'For every one billion particles of antimatter there were one billion and one particles of matter. And when the mutual annihilation was complete, one billionth remained – and that's our present universe.'**
>
> **Albert Einstein, 1879–1955**

Negative energy Creating an equation for the electron in 1928, British physicist Paul Dirac saw that it offered the possibility that electrons could have negative as well as positive energy. Just as the equation $x^2=4$ has the solutions $x=2$ and $x=-2$, Dirac had two ways of solving his problem: positive energy was expected, associated with a normal electron, but negative energy made no sense. But rather than ignore this confusing term, Dirac suggested that such particles might actually exist. This complementary state of matter is 'anti-'matter.

timeline

1928AD
Dirac derives the existence of antimatter

1932
Anderson detects the positron

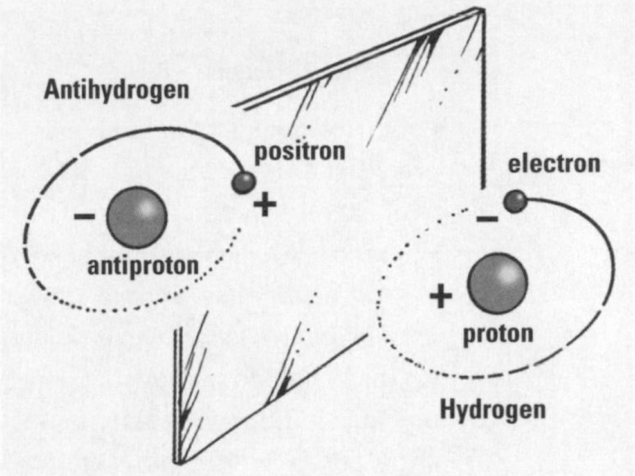

Antiparticles The hunt for antimatter began quickly. In 1932, Carl Anderson confirmed the existence of positrons experimentally. He was following the tracks of showers of particles produced by cosmic rays (energetic particles that crash into the atmosphere from space). He saw the track of a positively charged particle with the electron's mass, the positron. So antimatter was no longer just an abstract idea but was real.

Antihydrogen
positron
electron
antiproton
proton
Hydrogen

It took another two decades before the next antiparticle, the antiproton, was detected. Physicists built new particle-accelerating machines that used magnetic fields to increase the speeds of particles travelling through them. Such powerful beams of speeding protons produced enough energy to reveal the antiproton in 1955. Soon afterwards, the antineutron was also found.

With the antimatter equivalent building blocks in place, was it possible to build an anti-atom, or at least an anti-nucleus? The answer, shown in 1965, was yes. A heavy hydrogen (deuterium) anti-nucleus (an anti-deuteron), containing an antiproton and antineutron, was created by scientists at CERN in Europe and Brookhaven Laboratory in America. Tagging on a positron to an antiproton to make a hydrogen anti-atom (anti-hydrogen) took a little longer, but was achieved in 1995. Today experimenters are testing whether anti-hydrogen behaves in the same way as normal hydrogen.

On Earth, physicists can create antimatter in particle accelerators, such as those at CERN in Switzerland or Fermilab near Chicago. When the beams of particles and antiparticles meet, they annihilate each other in a flash of pure energy. Mass is converted to energy according to Einstein's $E=mc^2$ equation. So if you met your antimatter twin it might not be such a good idea to throw your arms around them.

1955
Antiprotons are detected

1965
The first anti-nucleus is produced

1995
Anti-hydrogen atoms are created

PAUL DIRAC 1902–84

Paul Dirac was a talented but shy British physicist. People joke that his vocabulary consisted of 'Yes', 'No', and 'I don't know'. He once said 'I was taught at school never to start a sentence without knowing the end of it.' What he lacked in verbosity he made up for in his mathematical ability. His PhD thesis is famous for being impressively short and powerful, presenting a new mathematical description of quantum mechanics. He partly unified the theories of quantum mechanics and relativity theory, but he also is remembered for his outstanding work on the magnetic monopole and in predicting antimatter. Awarded the 1933 Nobel prize, Dirac's first thought was to turn it down to avoid the publicity. But he gave in when told he would get even more publicity if he turned it down. Dirac did not invite his father to the ceremony, possibly because of strained relations after the suicide of Dirac's brother.

Universal asymmetries If antimatter were spread across the universe, these annihilation episodes would be occurring all the time. Matter and antimatter would gradually destroy each other in little explosions, mopping each other up. Because we don't see this, there cannot be much antimatter around. In fact normal matter is the only widespread form of particle we see, by a very large margin. So at the outset of the creation of the universe there must have been an imbalance such that more normal matter was created than its antimatter opposite.

'In science one tries to tell people, in such a way as to be understood by everyone, something that no one ever knew before. But in poetry, it's the exact opposite.'

Paul Dirac, 1902–84

Like all mirror images, particles and their antiparticles are related by different kinds of symmetry. One is time. Because of their negative energy, antiparticles are equivalent mathematically to normal particles moving backwards in time. So a positron can be thought of as an electron travelling from future to past. The next symmetry involves charges and other quantum properties, which are reversed, and is known as 'charge conjugation'. A third symmetry regards motion through space. Returning to Mach's principle, motions are generally unaffected if we change the direction of coordinates marking out the grid of space. A particle moving left to right looks the same as one moving right to left, or is unchanged whether spinning clockwise or anticlockwise. This 'parity' symmetry is true of most particles, but there are a few for which it does not always hold. Neutrinos exist in only one form, as a left-handed neutrino, spinning in one direction; there is no such thing as a right-handed neutrino. The converse is true for antineutrinos which are all right handed. So parity symmetry can sometimes be broken, although a combination of charge conjugation and parity is conserved, called charge–parity or CP symmetry for short.

> **‘The opposite of a correct statement is a false statement. But the opposite of a profound truth may well be another profound truth.’**
>
> **Niels Bohr**, 1885–1962

Just as chemists find that some molecules prefer to exist in one version, as a left-handed or right-handed structure, it is a major puzzle why the universe contains mostly matter and not antimatter. A tiny fraction – less than 0.01% – of the stuff in the universe is made of antimatter. But the universe also contains forms of energy, including a great many photons. So it is possible that a vast amount of both matter and antimatter was created in the big bang, but then most of it annihilated shortly after. Only the tip of the iceberg now remains. A minuscule imbalance in favour of matter would be enough to explain its dominance now. To do this, only 1 in every 10,000,000,000 (10^{10}) matter particles needed to survive a split second after the big bang, the remainder being annihilated. The leftover matter was likely preserved via a slight asymmetry from CP symmetry violation.

The particles that may have been involved in this asymmetry are a kind of heavy boson, called X bosons, that have yet to be found. These massive particles decay in a slightly imbalanced way to give a slight overproduction of matter. X bosons may also interact with protons and cause them to decay, which would be bad news as it means that all matter will eventually disappear into a mist of even finer particles. But the good news is that the timescale for this happening is *very* long. That we are here and no one has ever seen a proton decay means that protons are very stable and must live for at least 10^{17}–10^{35} years, or billions of billions of billions of years, hugely longer than the lifetime of the universe so far. But this does raise the possibility that if the universe gets really old, then even normal matter might, one day, disappear.

the condensed idea
Mirror image matter

34 Nuclear fission

The demonstration of nuclear fission is one of the great highs and lows of science. Its discovery marked a huge leap in our understanding of nuclear physics, and broke the dawn of atomic energy. But the umbrella of war meant this new technology was implemented almost immediately in nuclear weapons, devastating the Japanese cities of Hiroshima and Nagasaki and unleashing a proliferation problem that remains difficult to resolve.

At the start of the 20th century, the atom's inner world began to be revealed. Like a Russian doll, it contains many outer shells of electrons enveloping a hard kernel or nucleus. By the early 1930s, the nucleus itself was cracked, showing it to be a mix of positively charged protons and uncharged neutrons, both much heavier than the ephemeral electron, and bonded together by the strong nuclear force. Unlocking the energy glue of the nucleus became a holy grail of scientists.

Breakup The first successful attempt to split the nucleus occurred in 1932. Cockroft and Walton in Cambridge, England, fired very fast protons at metals. The metals changed their composition and released energy according to Einstein's $E=mc^2$. But these experiments needed more energy to be put into them than was created and so physicists didn't believe that it was possible to harness this energy for commercial use.

In 1938 German scientists Otto Hahn and Fritz Strassmann shot neutrons into the heavy element uranium, attempting to create new even heavier elements. What they found instead was that much lighter elements, some just half the mass of uranium, were given off. It was as if the nucleus

timeline

AD 1932
James Chadwick discovers the neutron

1938
Atomic fission is observed

sheared in half when bombarded by something less than half a percent of its mass; like a water melon splitting in two when hit by a cherry. Hahn wrote describing this to Lise Meitner, their exiled Austrian colleague who had just fled fascist Germany for Sweden. Meitner was equally puzzled and discussed it with her physicist nephew, Otto Frisch. Meitner and Frisch realized that energy would be released as the nucleus split because the two halves took up less energy overall. On his return to Denmark, Frisch could not contain his excitement and mentioned their idea to Niels Bohr. Embarking on a sea voyage to America, Bohr immediately set to work on an explanation, bringing the news to Italian Enrico Fermi at Columbia University.

‘. . . gradually we came to the idea that perhaps one should not think of the nucleus being cleaved in half as with a chisel but rather that perhaps there was something in Bohr’s idea that the nucleus was like a liquid drop.’

Otto Frisch, 1967

Meitner and Frisch published their paper ahead of Bohr, introducing the word ‘fission’ after the division of a biological cell. Back in New York, Fermi and Hungarian exile Léo Szilárd realized that this uranium reaction could produce spare neutrons that would produce more fission and so could go on to cause a nuclear chain reaction (a self-sustaining reaction). Fermi obtained the first chain reaction in 1942 at the University of Chicago, beneath the football stadium.

Nuclear power

Subcritical chain reactions can be kept stable and used for nuclear power plants. Boron control rods regulate the flow of neutrons through the uranium fuel by absorbing spare neutrons. In addition, coolant is needed to reduce the heat from the fission reactions. Water is most common, but pressurized water, helium gas, and liquid sodium can all be used. Today, France leads the world in using nuclear power, producing more than 70% of its total power compared with around 20% in the US or UK.

1942
The first chain reaction is obtained

1945
Atomic bombs are dropped on Japan

1951
Electricity is generated from nuclear power

Chain reaction Fellow physicist Arthur Compton remembered the day: 'On the balcony a dozen scientists were watching the instruments and handling the controls. Across the room was a large cubical pile of graphite and uranium blocks in which we hoped the atomic chain reaction would develop. Inserted into openings in this pile of blocks were control and safety rods. After a few preliminary tests, Fermi gave the order to withdraw the control rod another foot. We knew that that was going to be the real test. The geiger counters registering the neutrons from the reactor began to click faster and faster till their sound became a rattle. The reaction grew until there might be danger from the radiation up on the platform where we were standing. "Throw in the safety rods," came Fermi's order. The rattle of the counters fell to a slow series of clicks. For the first time, atomic power had been released. It had been controlled and stopped. Somebody handed Fermi a bottle of Italian wine and a little cheer went up.'

Manhattan project Szilárd was so concerned about German scientists copying their feat that he approached Albert Einstein and they presented a joint letter to warn President Roosevelt in 1939. However, not much happened until 1941 when physicists in the UK shared a calculation showing just how easy it was to build a nuclear weapon. That coincided with the Japanese attack at Pearl Harbor and Roosevelt soon began the US nuclear bomb project, known as the Manhattan project. It was led by

Nuclear waste

Fission reactors are efficient producers of energy, but they generate radioactive waste. The most toxic products include the remnants of the uranium fuels, that can remain radioactive for thousands of years, and heavier elements (such as plutonium) that can last for hundreds of thousands of years. These dangerous types of waste are made in only small quantities, but the extraction of uranium from its ore and other processes leaves a trail of lower level waste. How to dispose of this waste is a question that is still being decided worldwide.

Berkeley physicist Robert Oppenheimer from a remote and secret base at Los Alamos in New Mexico.

In the summer of 1942, Oppenheimer's team designed mechanisms for the bomb. To set up the chain reaction leading to the explosion a critical mass of uranium was needed, but it should be split up before detonation. Two techniques were favoured, a 'gun' mechanism where a lump of uranium was shot into another with conventional explosives to complete the critical mass, and an 'implosion' mechanism where conventional explosives caused a hollow sphere of uranium to implode onto a core of plutonium.

‘I thought this day would go down as a black day in the history of mankind . . . I was also aware of the fact that something had to be done if the Germans get the bomb before we have it . . . They had the people to do it . . . We had no choice, or we thought we had no choice.’

Léo Szilárd, 1898–1964

Uranium comes in two types, or isotopes, hosting different numbers of neutrons in their nuclei. The most common isotope, uranium-238 is ten times more common than the other, uranium-235. It is uranium-235 that is most effective for a fission bomb, so raw uranium is enriched in uranium-235. When uranium-238 receives a neutron it becomes plutonium-239. Plutonium-239 is unstable and its breakdown produces even more neutrons per gram, so mixing in plutonium can trigger the chain reaction readily. The gun method was used with enriched uranium to build the first type of bomb, called 'Little Boy'. The spherical implosion bomb type, including plutonium, was also built and named 'Fat Man'.

On 6 August 'Little Boy' was dropped on Hiroshima. Three days later, 'Fat Man' destroyed Nagasaki. Each bomb released the equivalent of about 20,000 tons of dynamite, killing 70,000–100,000 people immediately, and twice that eventually.

the condensed idea
Splitting the atom

35 Nuclear fusion

All the elements around us, including those in our bodies, are the product of nuclear fusion. Fusion powers stars like the Sun, within which all the elements heavier than hydrogen are cooked up. We really are made of stardust. If we can harness the stars' power on Earth, fusion could even be the key to unlimited clean energy.

Nuclear fusion is the merging together of light atomic nuclei to form heavier ones. When pressed together hard enough, hydrogen nuclei can merge to produce helium, giving off energy – a great deal of energy – in the process. Gradually, by building up heavier and heavier nuclei through a series of fusion reactions, all the elements that we see around us can be created from scratch.

> **'I ask you to look both ways. For the road to a knowledge of the stars leads through the atom; and important knowledge of the atom has been reached through the stars.'**
>
> Sir Arthur Eddington, 1928

Tight squeeze Fusing together even the lightest nuclei, such as hydrogen, is tremendously difficult. Enormous temperatures and pressures are needed, so fusion only happens naturally in extreme places, like the Sun and other stars. For two nuclei to merge, the forces that hold each one together must be overcome. Nuclei are made up of protons and neutrons locked together by the strong nuclear force. The strong force is dominant at the tiny scale of the nucleus, and is much weaker outside the nucleus. Because protons are positively charged, their electrical charges repel one another, so pushing each other apart slightly as well. But the strong force glue is more powerful so the nucleus holds together.

Because the strong nuclear force acts over such a short precise range, its combined strength is greater for small nuclei than for large ones. For a

timeline

AD 1920
Eddington applies the idea of fusion to stars

1932
Hydrogen fusion is demonstrated in the laboratory

weighty nucleus, such as uranium, with 238 nucleons, the mutual attraction will not be as strong between nucleons on opposite sides of the nucleus. The electric repulsive force, on the other hand, is still felt at larger separations and so becomes stronger for larger nuclei because it can span the whole nucleus. It is also boosted by the greater numbers of positive charges they contain. The net effect of this balance is that the energy needed to bind the nucleus together, averaged per nucleon, increases with atomic weight up to the elements nickel and iron, which are very stable, and then drops off again for larger nuclei. So fission of large nuclei happens relatively easily as they can be disrupted by a minor knock.

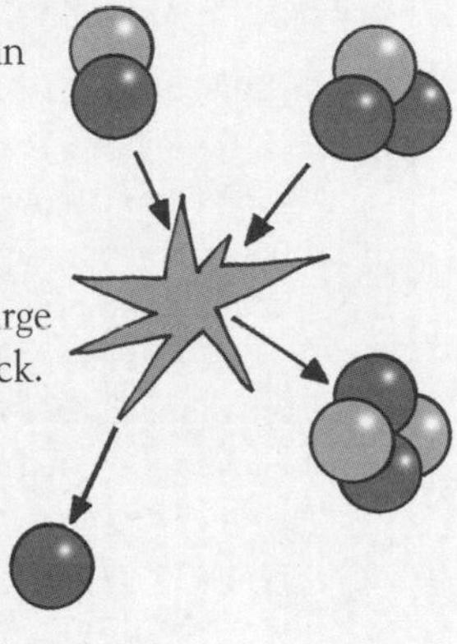

For fusion, the energy barrier to overcome is least for hydrogen isotopes that contain just a single proton. Hydrogen comes in three types: 'normal' hydrogen atoms contain one proton surrounded by a single electron; deuterium, or heavy hydrogen, has one proton, one electron and also one neutron; tritium has two neutrons added, so it is even heavier. The simplest fusion reaction therefore is the combination of hydrogen and deuterium to form tritium plus a lone neutron. Although it is the simplest, scorching temperatures of 800 million kelvins are needed to ignite even this reaction (which is why tritium is quite rare).

Fusion reactors On Earth, physicists are trying to replicate these extreme conditions in fusion reactors to generate power. However, they are decades off from achieving this in practice. Even the most advanced fusion machines take in more energy than they give out, by many orders of magnitude.

Fusion power is the holy grail of energy production. Compared with fission technology, fusion reactions are relatively clean and, should they work, efficient. Very few atoms are needed to produce huge amounts of energy (from Einstein's $E=mc^2$ equation), there is very little waste and certainly nothing as nasty as the ultraheavy elements that come out of fission reactors. Fusion power does not produce greenhouse gases either, promising

1939
Hans Bethe describes stellar fusion processes

1946/1954
Fred Hoyle explains the production of heavier elements

1957
Burbidge, Burbidge, Fowler, Hoyle publish a famous paper on nucleosynthesis

Cold fusion

In 1989, the scientific world was rocked by a controversial claim. Martin Fleischmann and Stanley Pons reported they had performed nuclear fusion, not in a huge reactor, but in a test tube. By firing electric current through a beaker of heavy water (whose hydrogen atoms are replaced by deuterium), the pair believed they had created energy via 'cold' fusion. They said their experiment gave out more energy than was put into it, due to fusion occurring. This caused uproar. Most scientists believed Fleischmann and Pons were mistaken in accounting for their energy budget, but even now this is not settled. Other disputed claims of lab-based fusion have occasionally cropped up. In 2002, Rudi Taleyarkhan proposed that fusion was behind so-called sono-luminescence, where bubbles in a fluid emit light when pulsed (and heated) rapidly by ultrasound waves. The jury is still out on whether fusion can really be made to work in a lab flask.

a self-contained, reliable source of energy assuming its fuel, hydrogen and deuterium, can be manufactured. But it is not perfect and will produce some radioactive by-products as neutrons are released in the main reactions and need to be mopped up.

At the high temperatures involved, controlling the scorching gases is the main difficulty, so although fusion has been achieved these monster machines only work for a few seconds at a time. To try and break through the next technology barrier, an international team of scientists is collaborating to build an even bigger fusion reactor in France, called the International Thermonuclear Experimental Reactor (ITER), that will test the feasibility of whether fusion can be made to work commercially.

Stardust Stars are nature's fusion reactors. German physicist Hans Bethe described how they shine by converting hydrogen nuclei (protons) into helium nuclei (two protons and two neutrons). Additional particles (positrons and neutrinos) are involved in the transfer, so that two of the original protons are turned into neutrons in the process.

Within stars, heavier elements are gradually built up in steps by fusion cookery, just like following a recipe. Larger and larger nuclei are constructed through a succession of 'burning' first hydrogen, then helium, then other elements lighter than iron and, eventually, elements heavier than iron. Stars like the Sun shine because they are mostly fusing hydrogen into helium and this proceeds slowly enough that heavy elements are made in only small quantities. In bigger stars this reaction is sped up by the involvement of the elements carbon, nitrogen and oxygen in further reactions. So more heavy elements are made more quickly. Once helium is present, carbon can be made from it (three helium-4 atoms fuse, via unstable beryllium-8). Once some carbon is made it can combine with helium to make oxygen, neon and magnesium. These slow transformations take most of the life of the star. Elements heavier than iron are made in slightly different reactions, gradually building sequences of nuclei right up the periodic table.

First stars Some of the first light elements were created not in stars but in the big bang fireball itself. At first the universe was so hot that not even atoms were stable. As it cooled hydrogen atoms condensed out first, along with a smattering of helium and lithium and a tiny amount of beryllium. These were the first ingredients for all stars and everything else. All elements heavier than this were created in and around stars and were then flung across space by exploding stars called supernovae. However, we still don't really understand how the first stars switched on. The very first star would not contain any heavy elements, just hydrogen, and so would not cool down quickly enough to collapse and switch on its fusion engine. The process of collapsing under gravity causes the hydrogen gas to heat up and swell, too much. Heavy elements can help it cool down by radiating light, so by the time that one generation of stars has existed and blown out all their by-products into space via supernovae, stars are easy to make. But forming the first star and its siblings quickly enough is still a challenge to the theorists.

‘We are bits of stellar matter that got cold by accident, bits of a star gone wrong.’

Sir Arthur Eddington, 1882–1944

Fusion is a fundamental power source across the universe. If we can tap it then our energy woes could be over. But it means harnessing the enormous power of the stars here on Earth, which isn't easy.

the condensed idea
Star power

36 Standard model

Protons, neutrons and electrons are just the tip of the particle physics iceberg. Protons and neutrons are made up of even smaller quarks, electrons are accompanied by neutrinos, and forces are mediated by a whole suite of bosons, including photons. The 'standard model' brings together the entire particle zoo in a single family tree.

To the Greeks, atoms were the smallest components of matter. It was not until the end of the 19th century that even smaller ingredients, first electrons and then protons and neutrons, were etched out of atoms. So are these three particles the ultimate building blocks of matter?

> 'Even if there is only one possible unified theory, it is just a set of rules and equations. What is it that breathes fire into the equations and makes a universe for them to describe?'
>
> Stephen Hawking, 1988

Well, no. Even protons and neutrons are grainy. They are made up of even tinier particles called quarks. And that's not all. Just as photons carry electromagnetic forces, a myriad of other particles transmit the other fundamental forces. Electrons are indivisible, as far as we know, but they are paired with near massless neutrinos. Particles also have their antimatter doppelgangers. This all sounds pretty complicated, and it is, but this plethora of particles can be understood in a single framework called the standard model of particle physics.

Excavation In the early 20th century physicists knew that matter was made up of protons, neutrons and electrons. Niels Bohr had described how, due to quantum theory, electrons arranged themselves in a series of shells around the nucleus, like the orbits of planets around the Sun. The properties of the nucleus were even stranger. Despite their repelling

timeline

c.400BC
Democritus proposes the idea of atoms

positive charges, nuclei could host tens of protons alongside neutrons compressed into a tiny hard kernel, bound by the precise strong nuclear force. But as more was learned from radioactivity about how nuclei broke apart (via fission) or joined together (via fusion), it became clear that more phenomena needed to be explained.

First, the burning of hydrogen into helium in the Sun, via fusion, implicates another particle, the neutrino, which transforms protons into neutrons. In 1930, the neutrino's existence was inferred to explain the decay of a neutron into a proton and electron – beta radioactive decay. The neutrino itself was not discovered until 1956, having virtually no mass. So, even in the 1930s there were many loose ends. Pulling on some of these dangling threads, in the 1940s and 50s other particles were sought and the collection grew.

Out of these searches evolved the standard model, which is a family tree of subatomic particles. There are three basic types of fundamental particle, 'hadrons' made of 'quarks', others called 'leptons' that include electrons, and then particles (bosons) that transmit forces, like photons. Each of the quarks and leptons has a corresponding antiparticle as well.

Quarks In the 1960s, by firing electrons at protons and neutrons physicists realized they hosted even smaller particles within them, called quarks. Quarks come in threes. They have three 'colours': red, blue and green. Just as electrons and protons carry electric charge, quarks carry 'colour charge', which is conserved when quarks change from one type to another. Colour charge is nothing to do with the visible colours of light – it is just physicists having to be inventive and finding an arbitrary way of naming the weird quantum properties of quarks.

Quarks

Quarks were so named after a phrase used in James Joyce's *Finnegans Wake* to describe the cries of seagulls. He wrote that they gave 'three quarks', or three cheers.

AD1930 Wolfgang Pauli predicts the existence of the neutrino

1956 Neutrinos are detected

1960s Quarks are proposed

1995 The top quark is found

> **'The creative element in the mind of man . . . emerges in as mysterious a fashion as those elementary particles which leap into momentary existence in great cyclotrons, only to vanish again like infinitesimal ghosts.'**
>
> **Sir Arthur Eddington, 1928**

Just as electric charges produce a force, so colour charges (quarks) can exert forces on one another. The colour force is transmitted by a force particle called a 'gluon'. The colour force gets stronger the further the quarks are apart, so they stick together as if held by an invisible elastic band. Because the colour force field tie is so strong, quarks cannot exist on their own and must always be locked together in combinations that are colour neutral overall (exhibiting no colour charge). Possibilities include threesomes called baryons, ('bary' means heavy) including normal protons and neutrons, or quark–antiquark pairs (called mesons).

As well as having colour charge, quarks come in 6 types, or 'flavours'. Three pairs make up each generation of increasing mass. The lightest are the 'up' and 'down' quarks; next come 'strange' and 'charm' quarks; finally, the top' and 'bottom' quarks are the heaviest pair. The up, charm and top quarks have electric charges $+^2/_3$ and the down, strange and bottom quarks have charge $-^1/_3$. So quarks have fractional electric charge, compared with +1 for protons or –1 for electrons. So three quarks are needed to make up a proton (two ups and a down) or a neutron (two downs and an up).

Leptons The second class of particles, the leptons, are related to and include electrons. Again there are three generations with increasing masses: electrons, muons and taus. Muons are 200 times heavier than an electron and taus 3700 times. Leptons all have single negative charge. They also have an associated particle called a neutrino (electron, muon- and tau-neutrino) that has no charge. Neutrinos have almost no mass and do not interact much with anything. They can travel right through the Earth without noticing, so are difficult to catch. All leptons have antiparticles.

Interactions Fundamental forces are mediated by the exchange of particles. Just as the electromagnetic wave can also be thought of as a stream of photons, the weak nuclear force can be thought of as being carried by W and Z particles while the strong nuclear force is transmitted via gluons. Like the photon, these other particles are bosons, which can all exist in the same quantum state simultaneously. Quarks and leptons are fermions and cannot.

Particle smashing How do we know about all these subatomic particles? In the second half of the 20th century physicists exposed the inner workings of atoms and particles using brute force, by smashing them apart. Particle physics has been described as taking an intricate Swiss watch and smashing it up with a hammer and looking at the shards to work out how it operates. Particle accelerators use giant magnets to accelerate particles to extremely high speeds and then smash those particle beams either into a target or into another oppositely directed beam. At modest speeds, the particles break apart a little and the lightest generations of particles are released. Because mass means energy, you need a higher-energy particle beam to release the later (heavier) generations of particles.

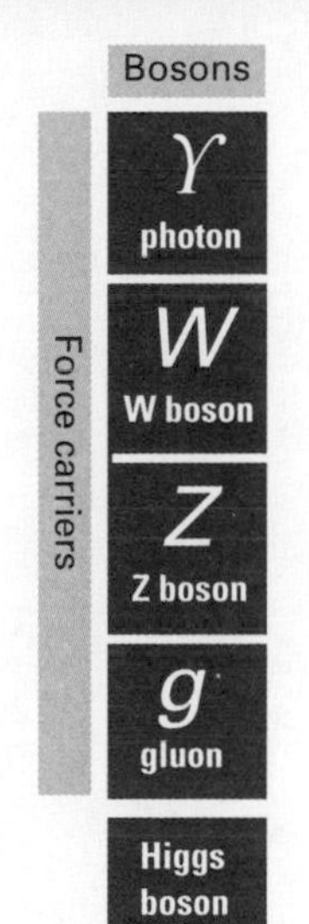

The particles produced in the atom smashers then need to be identified and particle physicists do this by photographing their tracks as they pass through a magnetic field. In the magnetic field, positive charged particles swerve one way and negative ones the other. The mass of the particle also dictates how fast it shoots through the detector and how much its path is curved by the magnetic field. So light particles barely curve and heavier particles may even spiral into loops. By mapping their characteristics in the detector, and comparing them with what they expect from their theories, particle physicists can tell what each particle is.

One thing that is not yet included in the standard model is gravity. The 'graviton', or gravity force carrying particle, has been postulated but only as an idea. Unlike light, there's no evidence yet for any graininess in gravity. Some physicists are trying to put gravity into the standard model in what would be a grand unified theory (GUT). But we are a long way off yet.

the condensed idea
All in the family

37 Feynman diagrams

Feynman diagrams are clever sketches that can be used as shorthand to work out complicated particle physics equations. Each particle interaction can be drawn as three arrows meeting in a point, two marking the incoming and outgoing particles and one showing the particle carrying the force. Adding many of them together, physicists can work out the probabilities of interactions occurring.

Feynman was so taken with his diagrams that he painted them on the side of his van. When someone asked why, he replied simply 'because I'm Richard Feynman'.

Richard Feynman was a charismatic Californian particle physicist, as famous for being a great lecturer and skilled bongo player as for his physics. He came up with a new symbolic language to describe particle interactions that, due to its simplicity, has been used ever since. As shorthand for complicated mathematical equations, Feynman simply drew arrows. One arrow represents each particle, one arriving and the other leaving, plus another squiggly one indicating the interaction. So every particle interaction can be shown as three arrows meeting at a point, or vertex. More complicated interactions can be built up from several of these shapes.

Feynman diagrams are more than just graphic tools. They not only help physicists to show the mechanisms by which subatomic particles interact but drawing them also helps physicists to calculate the probability that that interaction will take place.

Sketches Feynman diagrams depict particle interactions using a series of arrows to show the paths of the particles involved. The diagrams are usually

timeline

AD 1927
Work on quantum field theory begins

1940s
Quantum electrodynamics is developed

drawn so that time increases to the right, so incoming or outgoing electrons would be drawn as arrows pointing to the right. They are usually slanted to indicate movement. For antiparticles, because they are equivalent to real particles moving backwards in time, their arrows are shown pointing backwards, from right to left. Here are some examples.

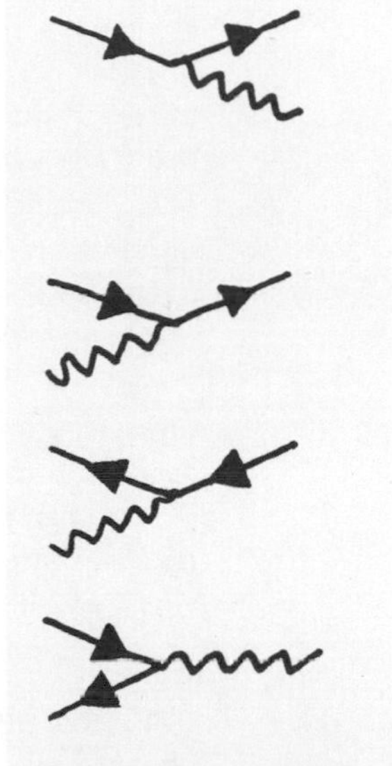

This diagram could represent an electron emitting a photon. The incoming electron (left hand arrow) experiences an electromagnetic interaction at the three-way intersection that produces another outgoing electron (right hand arrow) and a photon (wiggly line). The actual particle is not specified – just the mechanics of the interaction. It could equally well be a proton emitting a photon.

Here, the incoming electron, or other particle, absorbs a photon to produce a second more energetic electron.

Now the arrows are reversed so these must be antiparticles. This diagram might refer to an anti-electron, or positron (left arrow) absorbing a photon to produce another positron (right arrow).

And here, an electron and a positron combine and annihilate to give off a photon of pure energy.

Two or more triple vertices may be combined to show a sequence of events. Here a particle and antiparticle annihilate to create a photon that then decays into another particle–antiparticle pair.

These vertices can be used to represent many different types of interactions. They can be used for any particles, including quarks as well as leptons, and their corresponding interactions using the electromagnetic, weak or strong nuclear forces. They all follow some basic rules. Energy must be conserved, lines entering and leaving the diagram must be real particles (such as protons or neutrons and not free quarks that cannot exist

1945

Atomic bombs are researched and used

1975

Quantum chromodynamics is proposed

RICHARD FEYNMAN 1918–88

Richard Feynman was a brilliant and zany physicist. He obtained a perfect score on the Princeton student entrance exam and drew the attention of fellows such as Einstein. Joining the Manhattan project as a junior physicist Feynman claimed to have watched the test explosion directly, telling himself that it was safe to look though the glass of a windscreen because it would block ultraviolet rays. Bored, and stuck in the desert at Los Alamos, Feynman broke locks on filing cabinets by guessing the numbers that physicists chose for the codes, such as the natural log $e = 2.71828\ldots$ He left notes as a prank and his colleagues became worried that there was a spy in their midst. He also started drumming for entertainment, which gave him a reputation for being eccentric. After the war, Feynman moved to Caltech where he enjoyed teaching and was called 'the great explainer', authoring many books including the famous *Feynman Lectures on Physics*. He sat on the *Challenger* disaster panel that investigated the shuttle's explosion, and was typically outspoken. His work included developing QED, the physics of superfluids and the weak nuclear force. Later in his career he set out the beginnings of quantum computing and nanotechnology in a talk 'There's plenty of room at the bottom'. Feynman was an adventurous spirit and enjoyed travel. Being good with symbols, he even tried his hand at deciphering Mayan hieroglyphs. Fellow physicist Freeman Dyson once wrote that Feynman was 'half-genius, half-buffoon', but later revised this to 'all-genius, all-buffoon'.

in isolation) but intermediate stages can involve any subatomic particles and virtual particles as long as they are all mopped up into real particles by the end.

This picture describes beta radioactive decay. On the left is a neutron, composed of two 'down' quarks and one 'up' quark. It is transformed in the interaction into a proton, that consists of two up quarks and one down quark, plus an electron and antineutrino. Two interactions are involved. A down quark from the neutron changes to an up quark producing a W boson (shown as a wiggly line), the mediator of the weak nuclear force. The W boson then decays into an electron and an antineutrino. The W boson is not seen in the products of the interaction, but is involved in the intermediate stage.

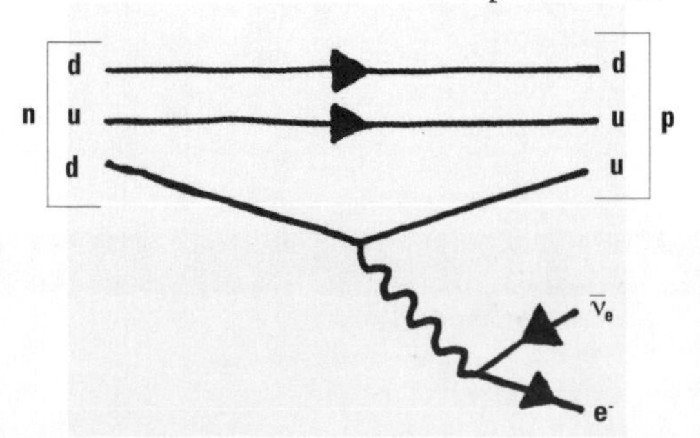

Probability These diagrams are not just convenient shorthand to visualize the interactions – they can also tell us how probable the interactions are. So they are powerful mathematical depictions of complicated equations too. In trying to work out how probable an interaction is you need to know how many ways there are of getting there. This is where the diagrams come into their own. By drawing all the different variations of the interactions, all the different ways you can get from the input to the output involving many interactions, you can work out the likelihoods of each one occurring by counting them up.

Similar sketches to Feynman's were used by particle physicist John Ellis, who called them penguin diagrams after a bet he had with his student in a bar that if he lost a game of darts he would have to use the word penguin in his next paper. His diagrams were arranged vertically on the page, and he thought that the diagrams looked a bit like penguins. The name stuck.

QED Feynman came up with his diagrams whilst developing quantum electrodynamics (QED) in the 1940s. The thinking behind QED is quite similar to Fermat's principle for the propagation of light: light follows all possible paths but it is the quickest path that is most probable, and where most of the light travels in phase. Applying a similar idea to electromagnetic fields, quantum field theory developed after 1927 and led on to QED.

QED describes the electromagnetic interactions, mediated by the exchange of photons, so it combined quantum mechanics with a description of the electric field and subatomic particles. It was in trying to work out the probabilities for all the possible interactions that Feynman came up with his graphic notation. After QED, physicists extended this picture to cover the colour force field of quarks, a theory called quantum chromodynamics, or QCD. And then QED was merged with the weak nuclear force into a combined 'electroweak' force.

the condensed idea
Three-pronged approach

38 The God particle

While walking in the Scottish Highlands in 1964, physicist Peter Higgs thought of a way to give particles their mass. He called this his 'one big idea'. Particles seem more massive because they are slowed while swimming through a force field, now known as the Higgs field. It is carried by the Higgs boson, referred to as the 'God particle' by Nobel laureate Leon Lederman.

Why does anything have a mass? A truck is heavy because it contains a lot of atoms each of which might itself be relatively heavy. Steel contains iron atoms and they fall far down the periodic table. But why is an atom heavy? It is mostly empty space after all. Why is a proton heavier than an electron, or a neutrino, or a photon?

Although the four fundamental forces, or interactions, were well known in the 1960s, they all relied on quite different mediating particles. Photons carry information in electromagnetic interactions, gluons link quarks by the strong nuclear force and the W and Z bosons carry weak nuclear forces. But photons have no mass, whereas the W and Z bosons are very massive particles, a hundred times as massive as the proton. Why are they so different? This discrepancy was particularly acute given that the theories of electromagnetic and weak forces could be combined, into an electroweak force. But this theory did not predict that the weak nuclear force particles, the W and Z bosons, should have a mass. They should be just like the massless photon. Any further combinations of fundamental forces, as attempted by the grand unified theory, also ran into the same problem. Force carriers should not have any mass. Why weren't they all like the photon?

timeline

AD 1687
Newton's *Principia* sets out equations for mass

Slow motion Higgs's big idea was to think of these force carriers as being slowed by passage through a background force field. Now called the Higgs field, it also operates by the transfer of bosons called Higgs bosons. Imagine dropping a bead into a glass. It will take longer to drop to the bottom if the glass is filled with water than if it is empty and filled with air. It is as if the bead is more massive when in water – it takes longer for gravity to pull it through the liquid. The same might apply to your legs if you walk through water – they feel heavy and your motion is slowed. The bead may be slowed even more if dropped into a glass of syrup, taking a while to sink. The Higgs field acts in a similar way, just like a viscous liquid. The Higgs force slows down the other force-carrying particles, effectively giving them a mass. It acts more strongly on the W and Z bosons than on photons, making them appear heavier.

> **'The obvious thing to do was to try it out on the simplest gauge theory of all, electrodynamics – to break its symmetry to see what really happens.'**
>
> **Peter Higgs, b.1929**

This Higgs field is quite similar to an electron moving through a crystal lattice of positively charged nuclei, such as a metal. The electron is slowed down a little because it is attracted by all the positive charges, so it appears to be more massive than in the absence of these ions. This is the electromagnetic force in action, mediated by photons. The Higgs field works similarly but Higgs bosons carry the force. You could also imagine it is like a film star walking into a cocktail party full of Higgs's. The star finds it hard to traverse the room because of all the social interactions slowing them down.

If the Higgs field gives the other force-carrier bosons mass, what is the mass of the Higgs boson? And how does it get its own mass? Isn't this a chicken and egg situation? Unfortunately the theories do not predict the mass of the Higgs boson itself, although they do predict the necessity for it

1964
Higgs has his insight into what gives particles mass

2007
The Large Hadron Collider at CERN is constructed

Symmetry breaking in magnets

At very high temperatures, all the atoms in a magnet are disordered, their inbuilt magnetic fields are all random and the material is not magnetic. But when the temperature drops below a certain point, called the Curie Temperature, the magnetic dipoles all align and produce an overall magnetic field.

within the standard model of particle physics. So physicists expect to see it, but they don't know how hard this will be or when it will appear (it has not been detected yet). Because of the ongoing search for particles with its properties, we know that its mass must be greater than the energies already reached experimentally. So it is very heavy, but we must wait to find out exactly how heavy.

Smoking gun The next machine that will take a good look for the Higgs particle is the Large Hadron Collider (LHC) at CERN in Switzerland. CERN, the Conseil Européen pour la Recherche Nucléaire (European Council for Nuclear Research) is a huge particle physics laboratory near Geneva. It houses rings of tunnels, the largest laid in a circle 27 km long, 100 m below ground. In the LHC, giant magnets accelerate protons forming a beam that curves around the track. They are constantly boosted as they go round, making them race faster and faster. Two opposing beams will be created and, when they are travelling at maximum speed, the beams will be fired into one another so that the speeding protons smash into each other head on. The huge energies produced will allow a whole range of massive particles to be released temporarily and recorded by detectors, along with their decay products if they are very short lived. It is the goal of the LHC to find the hint of the Higgs particle, buried amongst billions of other particle signatures. Physicists know what they are looking for, but it will still be hard to hunt it down. The Higgs may just appear, if the energies are high enough, for a fraction of a second, before disappearing into a cascade of other particles. So, rather than seeing the Higgs itself, the physicists will have to hunt for a smoking gun and then piece everything back together again to deduce its existence.

Symmetry breaking When might a Higgs boson appear? And how do we get from here to photons and other bosons? Because the Higgs boson must be very heavy it can only appear at extreme energies and, owing to Heisenberg's uncertainty principle (see page 104), only then for a very short time indeed. In the very early universe, theories suppose that all the forces were united together in one superforce. As the universe cooled the four fundamental forces dropped out, through a process called symmetry breaking.

Although symmetry breaking sounds quite a difficult thing to imagine, in fact it is quite simple. It marks the point where symmetry is removed from a system by one occurrence. An example is a round dinner table set with napkins and cutlery. It is symmetric in that it doesn't matter where you sit, the table looks the same. But if one person picks up their napkin the symmetry is lost – you can tell where you are relative to that position. So symmetry breaking has occurred. Just this one event can have knock on effects – it may mean that everyone else picks up their napkin on the left, to match the first event. If they had happened to take the napkin from the other side, then the opposite may have happened. But the pattern that follows is set up by a random event that triggered it. Similarly, as the universe cools, events cause the forces to decouple, one by one.

Even if scientists do not detect the Higgs boson with the LHC, it will be an interesting result. From neutrinos to the top quark, there are 14 orders of magnitude of mass that the standard model needs to explain. This is hard to do even with the Higgs boson, which is the missing ingredient. If we do find this God particle all will be well, but if it is not there then the standard model will need to be fixed. And that will require new physics. We think we know all the particles in the universe – the Higg's boson is the one remaining missing link.

the condensed idea
Swimming against the tide

39 String theory

Although most physicists are happy to work with the successful standard model, incomplete though it is, others are looking for new physics even before the standard model has been tested to destruction, or affirmation. In a modern twist on wave–particle duality, a group of physicists is trying to explain the patterns of fundamental particles by treating them not as hard spheres but as waves on a string. This idea has captured the imagination of the media and is known as string theory.

String theorists are not satisfied that fundamental particles, such as quarks, electrons and photons, are indivisible lumps of matter or energy. The patterns that give them a particular mass, charge or associated energy suggest another level of organization. These scientists believe that such patterns indicate deep harmonies. Each mass or energy quantum is a harmonic tone of the vibration of a tiny string. So particles can be thought of not as solid blobs but as vibrating strips or loops of string. In a way, this is a new take on Kepler's love of ideal geometric solids. It is as if the particles are all a pattern of notes that suggest a harmonic scale, played on a single string.

Vibrations In string theory, the strings are not as we know them on say a guitar. A guitar string vibrates in three dimensions of space, or perhaps we could approximate this to two if we imagine it is restricted to a plane along its length and up and down. But subatomic strings vibrate in just one dimension, rather than the zero dimensions of point-like particles. Their entire extent is not visible to us, but to do the mathematics, the scientists

timeline

AD 1921
The Kaluza–Klein theory is proposed for unifying electromagnetism and gravity

1970
Yoichiro Nambu describes the strong nuclear force using quantum mechanical strings

calculate the strings' vibrations over more dimensions, up to 10 or 11 of them. Our own world has three dimensions of space and one more of time. But string theorists think that there may be many more that we don't see, dimensions that are all curled up so we don't notice them. It is in these other worlds that the particle strings vibrate.

The strings may be open ended or closed loops, but they are otherwise all the same. So all the variety in fundamental particles arises just because of the pattern of vibration of the string, its harmonics, not the material of the string itself.

Offbeat idea String theory is an entirely mathematical idea. No one has ever seen a string, and no one has any idea how to know if one were there for sure. So there are no experiments that anyone has yet devised that could test whether the theory is true or not. It is said that there are as many string theories as there are string theorists. This puts the theory in an awkward position among scientists.

> **'Having those extra dimensions and therefore many ways the string can vibrate in many different directions turns out to be the key to being able to describe all the particles that we see.'**
> **Edward Witten, b.1951**

The philosopher Karl Popper thought that science proceeds mainly by falsification. You come up with an idea, test it with an experiment and if it is false then that rules something out, so you learn something new and science progresses. If the observation fits the model then you have not learned anything new. Because string theory is not fully developed it does not yet have any definite falsifiable hypotheses. Because there are so many variations of the theory, some scientists argue it is not real science. Arguments about whether it is useful or not fill the letters pages of journals and even newspapers, but string theorists feel that their quest is worthwhile.

mid-1970s	1984–6	1990s
A quantum gravity theory is obtained	The rapid expansion of string theory 'explains' all particles	Witten and others develop M-theory in 11 dimensions

M-theory

Strings are essentially lines. But in multidimensional space they are a limiting case of geometries that might include sheets and other many-dimensional shapes. This generalized theory is called M-theory. There is no single word that the 'M' stands for, but it could be membrane, or mystery. A particle moving through space scrawls out a line; if the point-like particle is dipped in ink, it traces out a linear path, that we call its world line. A string, say a loop, would trace out a cylinder. So we say it has a world sheet. Where these sheets intersect, and where the strings break and recombine, interactions occur. So M-theory is really a study of the shapes of all these sheets in 11-dimensional space.

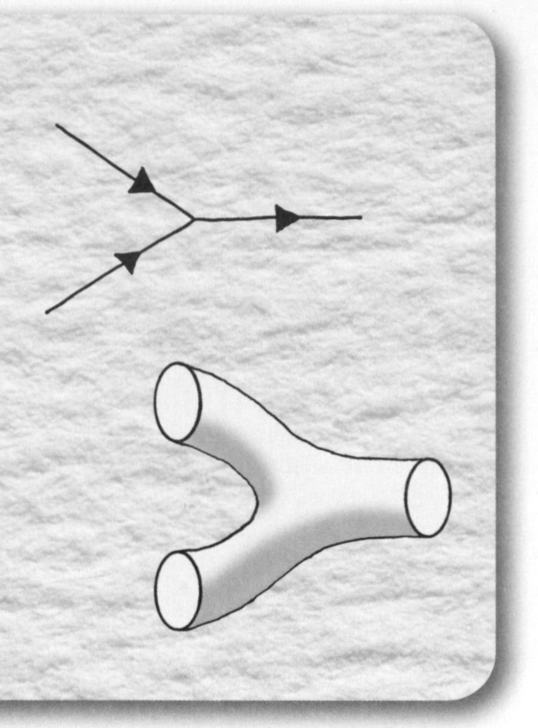

Theory of everything By trying to explain the whole zoo of particles and interactions within a single framework, string theory attempts to come close to a 'theory of everything', a single theory that unifies all four fundamental forces (electromagnetism, gravity and the strong and weak nuclear forces) and explains particle masses and all their properties. It would be a deep theory that underlies everything. Einstein originally tried to unify quantum theory and gravity in the 1940s, but he never succeeded, and nor has anyone since. Einstein was derided for his work as it was thought impossible and a waste of time. String theory brings gravity into the equations so its potential power draws people to pursue it. However, it is a long way from being precisely formulated let alone verified.

String theory arose as a novelty, owing to the beauty of its mathematics. In the 1920s Theodor Kaluza used harmonics as a different way to describe some unusual properties of particles. Physicists realized that this same mathematics could describe some quantum phenomena too. Essentially, the wavelike mathematics works well for both quantum mechanics and its extension into particle physics. This was then developed into early string theories. There are many variants, and it remains some way off from an all-encompassing theory.

‘I don’t like that they’re not calculating anything. I don’t like that they don’t check their ideas. I don’t like that for anything that disagrees with an experiment, they cook up an explanation – a fix-up to say, “Well, it still might be true.”’

Richard Feynman, 1918–88

A theory of everything is a goal of some physicists, who are generally reductionists and think that if you understand the building blocks then you can understand the whole world. If you understand an atom, built from vibrating strings, then you can infer all of chemistry, biology and so on. Other scientists find this whole attitude ridiculous. How can a knowledge of atoms tell you about social theory or evolution or taxes? Not everything can be scaled up simply. They think that such a theory describes the world as a pointless noise of subatomic interactions and is nihilistic and wrong. The reductionist viewpoint ignores evident macroscopic behaviour, such as the patterns of hurricanes or chaos and is described by physicist Steven Weinberg as ‘chilling and impersonal. It has to be accepted as it is, not because we like it, but because that is the way the world works.’

String theory, or rather theories, are still in a state of flux. No final theory has yet emerged, but this may take some time as physics has become so complicated there is a lot to include in it. Seeing the universe as the ringing of many harmonies has charm. But its adherents also sometimes verge on the dry side, being so engrossed in the fine detail that they diminish the significance of larger-scale patterns. Thus string theorists may stay on the sidelines until a stronger vision emerges. But given the nature of science, it is good that they are looking, and not in the usual places.

the condensed idea
Universal harmonies

40 Special relativity

Newton's laws of motion describe how most objects move, from cricket balls and cars to comets. But Albert Einstein showed in 1905 that strange effects happen when things move very quickly. Watching an object approach light speed, you'd see it become heavier, contract in length and age more slowly. That's because nothing can travel faster than the speed of light, so time and space themselves distort when approaching this universal speed limit.

Sound waves ring though air, but their vibrations cannot traverse empty space where there are no atoms. So it is true that 'in space no one can hear you scream'. But light is able to spread through empty space, as we know because we see the Sun and stars. Is space filled with a special medium, a sort of electric air, through which electromagnetic waves propagate? Physicists at the end of the 19th century thought so and believed that space was effused with a gas or 'ether' through which light radiates.

> **'The most incomprehensible thing about the world is that it is at all comprehensible.'**
>
> **Albert Einstein, 1879–1955**

Light speed In 1887, however, a famous experiment proved the ether did not exist. Because the Earth moves around the Sun, its position in space is always changing. If the ether were fixed then Albert Michelson and Edward Morley devised an ingenious experiment that would detect movement against it. They compared two beams of light travelling different paths, fired at right angles to one another and reflected back off identically faraway mirrors. Just as a swimmer takes less time to travel across a river from one bank to

timeline

AD 1881
Michelson and Morley are unable to verify the ether

1905
Einstein publishes the special theory of relativity

Twin paradox

Imagine if time dilation applied to humans. Well it could. If your identical twin was sent off into space on a rocket ship fast enough and for long enough, then they would age more slowly than you on Earth. On their return, they might find you to be elderly when they are still a sprightly youth. Although this seems impossible, it is not really a paradox because the space-faring twin would experience powerful forces that permit such a change to happen. Because of this time shift, events that appear simultaneous in one frame may not appear so in another. Just as time slows, so lengths contract also. The object or person moving at that speed would not notice either effect, it would just appear so to another viewer.

the other and back than to swim the same distance upstream against the current and downstream with it, they expected a similar result for light. The river current mimics the motion of the Earth through the ether. But there was no such difference – the light beams returned to their starting points at exactly the same time. No matter which direction the light travelled, and how the Earth was moving, the speed of light remained unchanged. Light's speed was unaffected by motion. The experiment proved the ether did not exist – but it took Einstein to realize this.

Just like Mach's principle (see page 4), this meant that there was no fixed background grid against which objects moved. Unlike water waves or sound waves, light appeared to always travel at the same speed. This was odd and quite different from our usual experience where velocities add together. If you are driving in a car at 50 km/h and another passes you at 65 km/h it is as if you are stationary and the other is travelling at 15 km/h past you. But even if you were rushing at hundreds of km/h light would still travel at the same speed. It is exactly 300 million metres per second whether you are shining a torch from your seat in a fast jet plane or the saddle of a bicycle.

1971

Time dilation is demonstrated by flying clocks in planes

> **'The introduction of a light-ether will prove to be superfluous since . . . neither will a space in absolute rest endowed with special properties be introduced nor will a velocity vector be associated with a point of empty space in which electromagnetic processes take place.'**
>
> **Albert Einstein, 1905**

It was this fixed speed of light that puzzled Albert Einstein in 1905, leading him to devise his theory of special relativity. Then an unknown Swiss patent clerk, Einstein worked out the equations from scratch in his spare moments. Special relativity was the biggest breakthrough since Newton and revolutionized physics. Einstein started with the assumption that the speed of light is a constant value, and appears the same for any observer no matter how fast they are moving. If the speed of light does not change then, reasoned Einstein, something else must change to compensate.

Space and time Following ideas developed by Edward Lorenz, George Fitzgerald and Henri Poincaré, Einstein showed that space and time must distort to accommodate the different viewpoints of observers travelling close to the speed of light. The three dimensions of space and one of time made up a four-dimensional world in which Einstein's vivid imagination worked. Speed is distance divided by time, so to prevent anything from exceeding the speed of light, distances must shrink and time slow down to compensate. So a rocket travelling away from you at near light speed looks shorter and experiences time more slowly than you do.

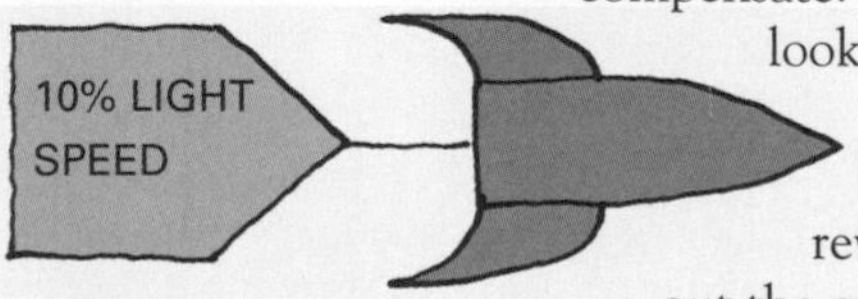

10% LIGHT SPEED

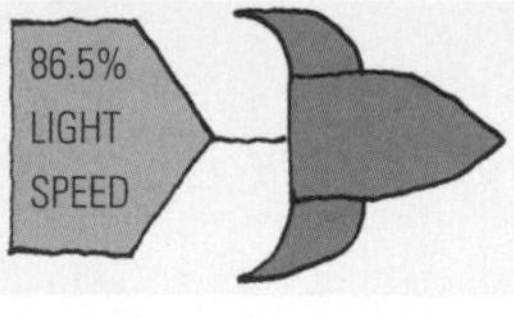

86.5% LIGHT SPEED

Einstein worked out how the laws of motion could be rewritten for observers travelling at different speeds. He ruled out the existence of a stationary frame of reference, such as the ether, and stated that all motion was relative with no privileged viewpoint. If you are sitting on a train and see the train next to you moving, you may not know whether it is your train or the other one pulling out. Moreover, even if you can see your train is stationary at the platform you cannot assume that you are immobile, just that you are not moving relative to that platform. We do not feel the motion of the Earth around the Sun; similarly, we never notice the Sun's path across our own Galaxy, or our Milky Way being pulled towards the huge Virgo cluster of galaxies beyond it. All that is experienced is relative motion, between you and the platform or the Earth spinning against the stars.

Einstein called these different viewpoints inertial frames. Inertial frames are spaces that move relative to one another at a constant speed, without experiencing accelerations or forces. So sitting in a car travelling at 50 km/h you are in one inertial frame, and you feel just the same as if you were in a train travelling at 100 km/h (another inertial frame) or a jet plane travelling at 500 km/h (yet another). Einstein stated that the laws of physics are the same in all inertial frames. If you dropped your pen in the car, train or plane, it would fall to the floor in the same way.

> **It is impossible to travel faster than the speed of light, and certainly not desirable, as one's hat keeps blowing off.**
>
> **Woody Allen**

Slower and heavier Turning next to understand relative motions near the speed of light, the maximum speed practically attainable by matter, Einstein predicted that time would slow down. Time dilation expressed the fact that clocks in different moving inertial frames may run at different speeds. This was proved in 1971 by sending four identical atomic clocks on scheduled flights twice around the world, two flying eastwards and two westwards. Comparing their times with a matched clock on the Earth's surface in the United States, the moving clocks had each lost a fraction of a second compared with the grounded clock, in agreement with Einstein's special relativity.

Another way that objects are prevented from passing the light speed barrier is that their mass grows, according to $E = mc^2$. An object would become infinitely large at light speed itself, making any further acceleration impossible. And anything with mass cannot reach the speed of light exactly, but only approach it, as the closer it gets the heavier and more difficult to accelerate it becomes. Light is made of mass-less photons so these are unaffected.

Einstein's special relativity was a radical departure from what had gone before. The equivalence of mass and energy was shocking, as were all the implications for time dilation and mass. Although Einstein was a scientific nobody when he published it, his ideas were read by Max Planck, and it is perhaps because of his adoption of Einstein's ideas that they became accepted and not sidelined. Planck saw the beauty in Einstein's equations, catapulting him to global fame.

the condensed idea
Motion is relative

41 General relativity

Incorporating gravity into his theory of special relativity, Einstein's theory of general relativity revolutionized our view of space and time. Going beyond Newton's laws, it opened up a universe of black holes, worm holes and gravitational lenses.

Imagine a person jumping off a tall building, or parachuting from a plane, being accelerated towards the ground by gravity. Albert Einstein realized that in this state of free fall they did not experience gravity. In other words they were weightless. Trainee astronauts today recreate the zero gravity conditions of space in just this way, by flying a passenger jet (attractively named the vomit comet) in a path that mimics a roller coaster. When the plane flies upwards the passengers are glued to their seats as they experience even stronger forces of gravity. But when the plane tips forwards and plummets downwards, they are released from gravity's pull and can float in the body of the aircraft.

Acceleration Einstein recognized that this acceleration was equivalent to the force of gravity. So, just as special relativity describes what happens in reference frames, or inertial frames, moving at some constant speed relative to one another, gravity was a consequence of being in a reference frame that is accelerating. He called this the happiest thought of his life.

Over the next few years Einstein explored the consequences. Talking through his ideas with trusted colleagues and using the latest mathematical formalisms to encapsulate them, he pieced together the full theory of

timeline

AD1687
Newton proposes his law of gravitation

1915
Einstein publishes the general theory of relativity

> ‘Time and space and gravitation have no separate existence from matter.’
>
> **Albert Einstein, 1915**

gravity that he called general relativity. The year 1915 when he published the work proved especially busy and almost immediately he revised it several times. His peers were astounded by his progress. The theory even produced bizarre testable predictions, including the idea that light could be bent by a gravitational field and also that Mercury's elliptical orbit would rotate slowly because of the gravity of the Sun.

Space–time In general relativity theory, the three dimensions of space and one of time are combined into a four-dimensional space–time grid, or metric. Light's speed is still fixed, and nothing can exceed it. When moving and accelerating, it is this space-time metric that distorts to maintain the fixed speed of light.

General relativity is best imagined by visualizing space–time as a rubber sheet stretched across a hollow table top. Objects with mass are like weighted balls placed on the sheet. They depress space–time around them. Imagine you place a ball representing the Earth on the sheet. It forms a depression in the rubber plane in which it sits. If you then threw in a smaller ball, say as an asteroid, it would roll down the slope towards the Earth. This shows how it feels gravity. If the smaller ball was moving fast enough and the Earth's dip was deep enough, then just as a daredevil cyclist can ride around an inclined track, that body would maintain a moon-like circular orbit. You can think of the whole universe as a giant rubber sheet. Every one of the planets and stars and galaxies causes a

1919
Eclipse observations verify Einstein's theory

1960s
Evidence for black holes is seen in space

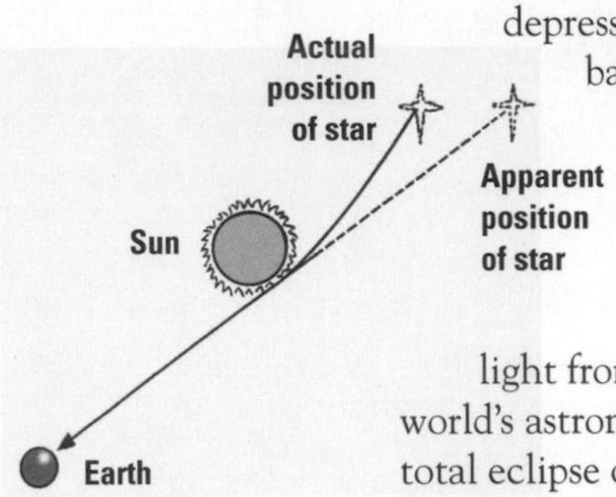

Actual position of star

Apparent position of star

Sun

Earth

depression that can attract or deflect passing smaller objects, like balls rolling over the contours of a golf course.

Einstein understood that, because of this warping of space time, light would be deflected if it passed near a massive body, such as the Sun. He predicted that the position of a star observed just behind the Sun would shift a little because light from it is bent as it passes the Sun's mass. On 29 May 1919 the world's astronomers gathered to test Einstein's predictions by observing a total eclipse of the Sun. It proved one of his greatest moments, showing that the theory some thought crazy was in fact close to the truth.

Warps and holes The bending of light rays has now been confirmed with light that has travelled right across the universe. Light from very distant galaxies clearly flexes when it passes a very massive region such as a giant cluster of galaxies or a really big galaxy. The background dot of light is smeared out into an arc. Because this mimics a lens the effect is known as gravitational lensing. If the background galaxy is sitting right behind the heavy intervening object then its light is smeared out into a complete circle, called an Einstein ring. Many beautiful photographs of this spectacle have been taken with the Hubble Space Telescope.

Gravity waves

Another aspect of general relativity is that waves can be set up in the space–time sheet. Gravitational waves can radiate, especially from black holes and dense spinning compact stars like pulsars. Astronomers have seen pulsars' spin decreasing so they expect that this energy will have been lost to gravity waves, but the waves have not yet been detected. Physicists are building giant detectors on Earth and in space that use the expected rocking of extremely long laser beams to spot the waves as they pass by. If gravity waves were detected then this would be another coup for Einstein's general relativity theory.

> ‘We shall therefore assume the complete physical equivalence of a gravitational field and the corresponding acceleration of the reference frame. This assumption extends the principle of relativity to the case of uniformly accelerated motion of the reference frame.’
>
> **Albert Einstein, 1907**

Einstein’s theory of general relativity is now widely applied to modelling the whole universe. Space–time can be thought of like a landscape, complete with hills, valleys and pot holes. General relativity has lived up to all observational tests so far. The regions where it is tested most are where gravity is extremely strong, or perhaps very weak.

Black holes (see page 168) are extremely deep wells in the space–time sheet. They are so deep and steep that anything that comes close enough can fall in, even light. They mark holes, or singularities, in space–time. Space–time may also warp into worm holes, or tubes, but no one has actually seen such a thing yet.

At the other end of the scale, where gravity is very weak it might be expected to break up eventually into tiny quanta, similar to light that is made up of individual photon building blocks. But no one has yet seen any graininess in gravity. Quantum theories of gravity are being developed but, without evidence to back it up, the unification of quantum theory and gravity is elusive. This hope occupied Einstein for the rest of his career but even he did not manage it and the challenge still stands.

the condensed idea
Warped space–time

42 Black holes

Falling into a black hole would not be pleasant, having your limbs torn asunder and all the while appearing to your friends to be frozen in time just as you fell in. Black holes were first imagined as frozen stars whose escape velocity exceeds that of light, but are now considered as holes or 'singularities' in Einstein's space–time sheet. Not just imaginary, giant black holes populate the centres of galaxies, including our own, and smaller ones punctuate space as the ghosts of dead stars.

If you throw a ball up in the air, it reaches a certain height and then falls back down. The faster you fling it the higher it goes. If you hurled it fast enough it would escape the Earth's gravity and whiz off into space. The speed that you need to reach to do this, called the 'escape velocity', is 11 km/s (or about 25,000 mph). A rocket needs to attain this speed if it is to escape the Earth. The escape velocity is lower if you are standing on the smaller Moon: 2.4 km/s would do. But if you were standing on a more massive planet then the escape velocity rises. If that planet was heavy enough, then the escape velocity could reach or exceed the speed of light itself, and so not even light could escape its gravitational pull. Such an object, that is so massive and dense that not even light can escape it, is called a black hole.

> **'God not only plays dice, but also sometimes throws them where they cannot be seen.'**
>
> **Stephen Hawking, 1977**

Event horizon The black hole idea was developed in the 18th century by geologist John Michell and mathematician Pierre-Simon Laplace. Later, after Einstein had proposed his relativity theories, Karl Schwarzschild worked out what a black hole would look like. In Einstein's theory of

timeline

AD1784
Michell deduces the possiblity of 'dark stars'

1930s
The existence of froze stars is predicted

general relativity, space and time are linked and behave together like a vast rubber sheet. Gravity distorts the sheet according to an object's mass. A heavy planet rests in a dip in space–time, and its gravitational pull is equivalent to the force felt as you roll into the dip, perhaps warping your path or even pulling you into orbit.

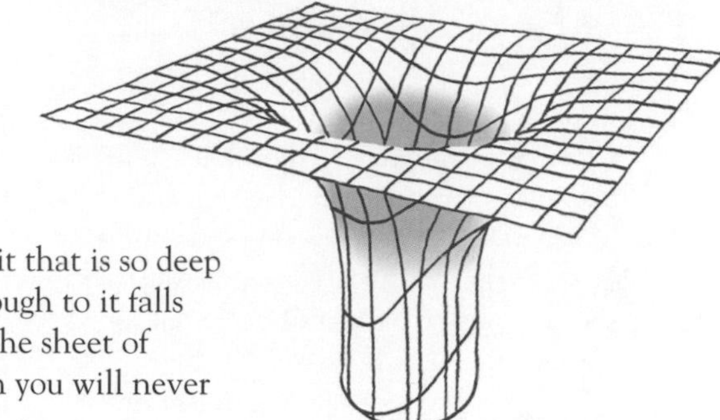

So what then is a black hole? It would be a pit that is so deep and steep that anything that comes close enough to it falls straight in and cannot return. It is a hole in the sheet of space–time, like a basketball net (from which you will never get your ball back).

If you pass far from a black hole, your path might curve towards it, but you needn't fall in. But if you pass too close to it, then you will spiral in. The same fate would even befall a photon of light. The critical distance that borders these two outcomes is called the 'event horizon'. Anything that falls within the event horizon plummets into the black hole, including light.

Falling into a black hole has been described as being 'spaghetti-fied'. Because the sides are so steep, there is a very strong gravity gradient within the black hole. If you were to fall into one feet first, and let's hope you never do, then your feet would be pulled more than your head and so your body would be stretched like being on a rack. Add to that any spinning motion and you would be pulled out like chewing gum into a scramble of spaghetti. Not a nice way to go. Some scientists have worried about trying to protect an unlucky person who might accidentally stumble into a black hole. One way you could protect yourself, apparently, is to don a leaden life-saver ring. If the ring was heavy and dense enough, it would counteract the gravity gradient and preserve your shape, and life.

1965
Quasars are discovered

1967
Wheeler renames frozen stars as black holes

1970s
Hawking proposes black holes evaporate

Evaporation

Strange as it may sound, black holes eventually evaporate. In the 1970s, Stephen Hawking suggested that black holes are not completely black but radiate particles due to quantum effects. Mass is gradually lost in this way and so the black hole shrinks until it disappears. The black hole's energy continually creates pairs of particles and their corresponding antiparticles. If this happens near the event horizon then sometimes one of the particles might escape even if the other falls in. To an outside eye the black hole seems to emit particles, called Hawking radiation. This radiated energy then causes the hole to diminish. This idea is still based in theory, and no one really knows what happens to a black hole. The fact that they are relatively common suggests that this process takes a long time, so black holes hang around.

Frozen stars The name 'black hole' was coined in 1967 by John Wheeler as a catchier alternative to describe a frozen star. Frozen stars were predicted in the 1930s by Einstein and Schwarzschild's theories. Because of the weird behaviour of space and time close to the event horizon, glowing matter falling in would seem to slow down as it does so, due to the light waves taking longer and longer to reach an observer looking on. As it passes the event horizon, this outside observer sees time actually stop so that the material appears to be frozen at the time it crosses the horizon. Hence, frozen stars, frozen in time just at the point of collapsing into the event horizon, were predicted. Astrophysicist Subrahmanyan Chandrasekhar predicted that stars more than 1.4 times the Sun's mass would ultimately collapse into a black hole; however, due to the Pauli exclusion principle (see page 120) we now know that white dwarfs and neutron stars will prop themselves up by quantum pressure, so black holes need more than 3 times the Sun's mass to form. Evidence of these frozen stars or black holes was not discovered until the 1960s.

If black holes suck in light, how can we see them to know they are there? There are two ways. First, you can spot them because of the way they pull other objects towards them. And second, as gas falls into them it can heat up and glow before it disappears. The first method has been used to identify a black hole lurking in the centre of our own Galaxy. Stars that pass close to it have been seen to whip past it and be flung out on elongated orbits. The Milky Way's black hole has a mass of a million Suns, squashed into a region of radius just 10 million kilometres (30 light seconds) or so. Black holes that lie in galaxies are called supermassive black holes. We don't know how they formed, but they seem to affect how galaxies grow so might have been there from day one or, perhaps, grew from millions of stars collapsing into one spot.

The second way to see a black hole is by the light coming from hot gas that is fired up as it falls in. Quasars, the most luminous things in the universe, shine due to gas being sucked into supermassive black holes in the centres of distant galaxies. Smaller black holes, just a few solar masses, can also be identified by X-rays shining from gas falling towards them.

‘The black holes of nature are the most perfect macroscopic objects there are in the universe: the only elements in their construction are our concepts of space and time.’

Subrahmanyan Chandrasekhar, 1983

Wormholes What lies at the bottom of a black hole in the space–time sheet? Supposedly they just end in a sharp point, or truly are holes, punctures in the sheet. But theorists have asked what might happen if they joined another hole. You can imagine that two nearby black holes might appear as long tubes dangling from the space–time sheet. If the tubes were joined together, then you could imagine a tube or wormhole being formed between the two mouths of the black holes. Armed with your ‘life-saver’ you might be able to jump into one black hole and pop out of another. This idea has been used a lot in science fiction for transport across time and space. Perhaps the wormhole could flow through to an entirely different universe. The possibilities for rewiring the universe are endless, but don't forget your life-saver ring.

the condensed idea
Light traps

43 Olbers' paradox

Why is the night sky dark? If the universe were endless and had existed for ever then it should be as bright as the Sun, yet it is not. Looking up at the night sky you are viewing the entire history of the universe. The limited number of stars is real and implies that the universe has a limited size and age. Olbers' paradox paved the way for modern cosmology and the big bang model.

You might think that mapping the entire universe and viewing its history would be difficult and call for expensive satellites in space, huge telescopes on remote mountaintops, or a brain like Einstein's. But in fact if you go out on a clear night you can make an observation that is every bit as profound as general relativity. The night sky is dark. Although this is something we take for granted, the fact that it is dark and not as bright as the Sun tells us a lot about our universe.

Star light star bright If the universe were infinitely big, extending for ever in all directions, then in every direction we look we would eventually see a star. Every sight line would end on a star's surface. Going further away from the Earth, more and more stars would fill space. It is like looking through a forest of trees – nearby you can distinguish individual trunks, appearing larger the closer they are, but more and more distant trees fill your view. So, if the forest was really big, you could not see the landscape beyond. This is what would happen if the universe were infinitely big. Even though the stars are more widely spaced than the trees, eventually there would be enough of them to block the entire view.

AD1610

Kepler notes the night sky is dark

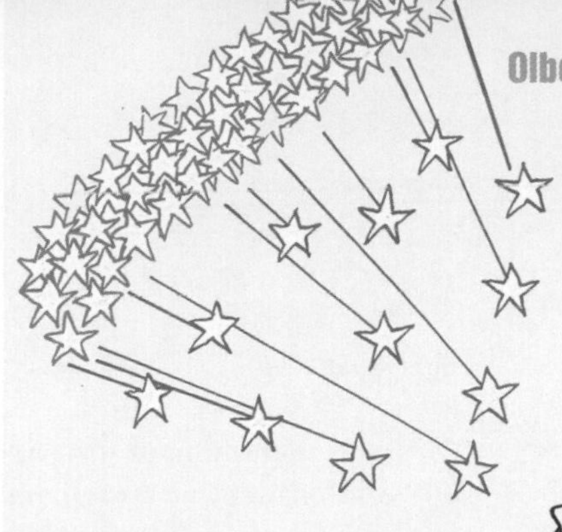

If all the stars were like the Sun, then every point of sky would be filled with star light. Even though a single star far away is faint, there are more stars at that distance. If you add up all the light from those stars they provide as much light as the Sun, so the entire night sky should be as bright as the Sun.

Obviously this is not so. The paradox of the dark night sky was noted by Johannes Kepler in the 17th century, but only formulated in 1823 by German astronomer Heinrich Olbers. The solutions to the paradox are profound. There are several explanations, and each one has elements of truth that are now understood and adopted by modern astronomers. Nevertheless, it is amazing that such a simple observation can tell us so much.

Dark skies

The beauty of the dark night sky is becoming harder and harder to see due to the glow of lights from our cities. On clear nights throughout history people have been able to look upward and see a brightly lit backbone of stars, stretched across the heavens. This was christened the Milky Way, and we now know that when we gaze at it we are looking towards the central plane of our Galaxy. Even in large cities 50 years ago it was possible to see the brightest stars and the Milky Way's swath, but nowadays hardly any stars are visible from towns and even the countryside views of the heavens are washed out by yellow smog. The vista that has inspired generations before us is becoming obscured. Sodium street lights are the main culprit, especially ones that waste light by shining upwards as well as down. Groups worldwide, such as the International Dark-Sky association, which includes astronomers, are now campaigning for curbs on light pollution so that our view out to the universe is preserved.

1832

Olbers formulates the paradox named after him

1912

Vesto Slipher measures the redshifts of galaxies

Eureka!

Edgar Allan Poe, in his 1848 prose poem *Eureka*, observed:

'Were the succession of stars endless, then the background of the sky would present us an uniform luminosity, like that displayed by the Galaxy – since there could be absolutely no point, in all that background, at which would not exist a star. The only mode, therefore, in which, under such a state of affairs, we could comprehend the voids which our telescopes find in innumerable directions, would be by supposing the distance of the invisible background so immense that no ray from it has yet been able to reach us at all.'

End in sight The first explanation is that the universe is not infinitely big. It must stop somewhere. So there must be a limited number of stars in it and not all sightlines will find a star. Similarly, standing near the edge of the forest or in a small wood you can see the sky beyond.

Another explanation could be that the more distant stars are fewer in number, so they do not add together to give as much light. Because light travels with a precise speed, the light from distant stars takes longer to reach us than from nearby stars. It takes 8 minutes for light to reach us from the Sun but 4 years for light from the next nearest star, Alpha Centauri to arrive, and as much as 100 thousand years for light to reach us from stars on the other side of our own Galaxy. Light from the next nearest galaxy, Andromeda, takes 2 million years to reach us; it is the most distant object we can see with the naked eye. So as we peer further into the universe, we are looking back in time and distant stars look younger than the ones nearby. This could help us with Olbers' paradox if those youthful stars eventually become rarer than Sun-like stars nearby. Stars like the Sun live for about 10 billion years (bigger ones live for shorter times and smaller ones for longer), so the fact that stars have a finite lifetime could also explain the paradox. Stars cease to exist earlier than a certain time because they have not been born yet. So stars have not existed for ever.

Making distant stars fainter than the Sun is also possible through redshift. The expansion of the universe stretches light wavelengths causing the light from distant stars to appear redder. So stars a long way away will look a little cooler than stars nearby. This could also restrict the amount of light reaching us from the outermost parts of the universe.

Wackier ideas have been put forward such as the distant light being blocked out, by soot from alien civilizations, iron needles or weird grey dust. But any absorbed light would be re-radiated as heat and so would turn up elsewhere in the spectrum. Astronomers have checked the light in the night sky at all wavelengths, from radio waves to gamma rays, and they have seen no sign that the visible star light is blocked.

Middle of the road universe So, the simple observation that the night sky is dark tells us that the universe is not infinite. It has only existed for a limited amount of time, it is restricted in size, and the stars in it have not existed forever.

Modern cosmology is based on these ideas. The oldest stars we see are around 13 billion years old, so we know the universe must have been formed before this time. Olbers' paradox suggests it cannot be very much ahead of this or we would expect to see many previous generations of stars and we do not.

Distant galaxies of stars are indeed redder than nearby ones, due to redshift, making them harder to see with optical telescopes and confirming that the universe is expanding. The most distant galaxies known today are so red they become invisible and can only be picked up at infrared wavelengths. So all this evidence supports the idea of the big bang, such that the universe grew out of a vast explosion some 14 billion years ago.

the condensed idea
Our finite universe

44 Hubble's law

Edwin Hubble was first to realize that galaxies beyond our own are all moving away from us together. The further away they are, the faster they recede, following Hubble's law. This galactic diaspora formed the first evidence that the universe is expanding, an astounding finding that changed our view of our entire universe and its destiny.

Copernicus's deduction in the 16th century that the Earth goes around the Sun caused major consternation. Humans no longer inhabited the exact centre of the cosmos. But in the 1920s, American astronomer Edwin Hubble made telescope measurements that were even more unsettling. He showed the entire universe was not static but expanding. Hubble mapped out the distances to other galaxies and their relative speeds compared with our Milky Way; he found that they were all hurtling away from us. We were so cosmically unpopular that only a few close neighbours were inching towards us. The more distant the galaxy, the faster it was receding, with a speed proportional to its distance away (Hubble's law). The ratio of speed to distance is always the same number and is called the Hubble constant. Astronomers today have measured its value to be close to 75 kilometres per second per megaparsec (a megaparsec, or a million parsecs, is equivalent to 3,262,000 light years or 3×10^{22} m). So galaxies continually recede from us by this amount.

> **'The history of astronomy is a history of receding horizons.'**
>
> **Edwin Hubble, 1938**

The great debate Before the 20th century, astronomers barely understood our own Galaxy, the Milky Way. They had measured hundreds of stars within it but also noted it was marked with many faint smudges, called nebulae. Some of these nebulae were gas clouds associated with the

timeline

AD 1918 Vesto Slipher measures redshifts of nebulae

1920 Shapley and Curtis debate the size of the Milky Way

births and deaths of stars. But some looked different. Some had spiral or oval shapes that suggested they were more regular than a cloud.

In 1920 two famous astronomers held a debate on the origin of these smudges. Harlow Shapley argued that everything in the sky was part of the Milky Way, which constituted the entire universe. On the other side, Heber Curtis proposed that some of these nebulae were separate 'island universes' or external 'universes' outside our own Milky Way. The term 'galaxy' was coined only later to describe these nebulous universes. Both astronomers cited evidence to back up their own idea, and the debate was not settled on the day. Later work by Hubble showed that Curtis's view was correct. These spiral nebulae really were external galaxies and did not lie within the Milky Way. The universe had suddenly opened up into a vast canvas.

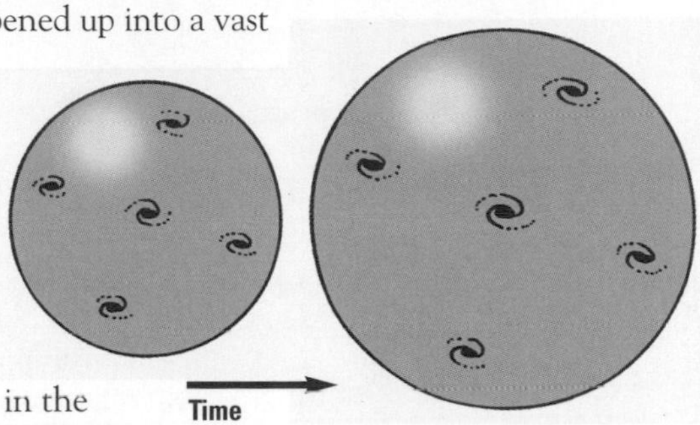

Time

Flying apart Hubble used the 100 inch Hooker Telescope at Mount Wilson to measure the light from flickering stars in the Andromeda nebula, now known to be a spiral galaxy very similar to the Milky Way and also a sibling in the group of galaxies associated with us. These flickering stars are called Cepheid variable stars, after the prototype star found in the constellation Cepheus, and are even now invaluable probes of distance. The amount and timing of flickering scales with the intrinsic brightness of the star, so once you know how its light varies you know how bright it is. Knowing its brightness you can then work out how far away it is because it is dimmed when placed at a distance. It is analogous to seeing a light bulb placed a distance away, reading that its power is 100 Watts, and then working out how far away it is by comparing its brightness with a 100 Watt bulb in front of you.

In this way Hubble measured the distance to the Andromeda galaxy. It was much further away than the size of our Milky Way, as given by

1922
Alexander Friedmann publishes the big bang model

1924
Cepheid variable stars are discovered

1929
Hubble and Milton Humason discover Hubble's law

Hubble Space Telescope

The Hubble Space Telescope is surely the most popular satellite observatory ever. Its stunning photographs of nebulae, distant galaxies and disks around stars have graced the front pages of many newspapers for almost 20 years. Launched in 1990 from the space shuttle *Discovery*, the spacecraft is about the size of a double-decker bus, 13 m long, 4 m across and weighing 11,000 kg. It carries an astronomical telescope whose mirror is 2.4 m across and a suite of cameras and electronic detectors that are able to take crystal clear images, in visible and ultraviolet light and the infrared. Hubble's power lies in the fact it is located above the atmosphere – so its photographs are not blurred. Now getting old, Hubble's fate is uncertain. NASA may upgrade its instruments but that would require a manned shuttle crew, or it may terminate its programme and either rescue the craft for posterity or crash it safely into the ocean.

Shapley, so it must lie outside. This simple fact was revolutionary. It meant that the universe was vast, and filled with other galaxies just like the Milky Way. If putting the Sun at the centre of the universe annoyed the church and humans' sensibility then demoting the Milky Way to just one in millions of other galaxies was a bigger blow to the human ego.

'We find them smaller and fainter, in constantly increasing numbers, and we know that we are reaching into space, farther and farther, until, with the faintest nebulae that can be detected with the greatest telescopes, we arrive at the frontier of the known universe.'

Edwin Hubble, 1938

Hubble then set about mapping distances to many other galaxies. He also found that the light from them was mostly redshifted by an amount that scaled with distance. The redshift is similar to the Doppler shift of a speeding object (see page 76). Finding that frequencies of light, such as atomic transitions of hydrogen, were all redder than expected meant that these galaxies were all rushing away from us, like many ambulance sirens falling off in tone as they speed away. It was very strange that all the galaxies were rushing away, with only 'local' ones moving towards us. The further away you looked, the faster they receded. Hubble saw that the galaxies weren't simply receding from us, which would have made our place in the universe very privileged indeed. Instead, they were all hurtling away from each other. Hubble concluded that the universe itself was expanding, being inflated like a giant balloon. The galaxies are like spots on the balloon, getting further apart from one another as more air is added.

How far how fast? Even today astronomers use Cepheid variable stars to map out the local universe's expansion. Measuring the Hubble constant accurately has been a major goal. To do so you need to know how far away something is and its speed or redshift. Redshifts are straightforward to measure from atomic spectra. The frequency of a particular atomic transition in star light can be checked against its known wavelength in the laboratory; the difference gives its redshift. Distances are harder to determine, because you need to observe something in the distant galaxy either whose true length is known or whose true brightness is known, a 'standard candle'.

There are a variety of methods for inferring astronomical distances. Cepheid stars work for nearby galaxies when you can separate the individual stars. But further away other techniques are needed. All the different techniques can be tied together one by one to build up a giant measuring rod, or 'distance ladder'. But because each method comes with peculiarities there are still many uncertainties in the accuracy of the extended ladder.

The Hubble constant is now known to an accuracy of about 10%, thanks largely to observations of galaxies with the Hubble Space Telescope and the cosmic microwave background radiation. The expansion of the universe began in the big bang, the explosion that created the universe, and galaxies have been flying apart ever since then. Hubble's law sets a limit on the age of the universe. Because it is continuously expanding, if you trace back the expansion to the beginning point, you can work out how long ago that was. It turns out to be around 14 billion years. This expansion rate is fortunately not enough to break apart the universe. The cosmos instead is finely balanced, in between completely blowing apart and containing enough mass to collapse back in on itself eventually.

the condensed idea
The expanding universe

45 The big bang

The birth of the universe in a phenomenal explosion created all space, matter and time as we know it. Predicted from the mathematics of general relativity, we see evidence for the big bang in the rush of galaxies away from our own, the quantities of light elements in the universe and the microwave glow that fills the sky.

The big bang is the ultimate explosion – the birth of the universe. Looking around us today, we see signs that our universe is expanding and infer it must have been smaller, and hotter, in the past. Taking this to its logical conclusion means that the entire cosmos could have originated from a single point. At the moment of ignition, space and time and matter were all created together in a cosmic fireball. Very gradually, over 14 billion years, this hot, dense cloud swelled and cooled. Eventually it fragmented to produce the stars and galaxies that dot the heavens today.

It's no joke The 'big bang' phrase itself was actually coined in ridicule. The eminent British astronomer Fred Hoyle thought it preposterous that the whole universe grew from a single seed. In a series of lectures first broadcast in 1949 he derided as far-fetched the proposition of Belgian mathematician Georges Lemaître who found such a solution in Einstein's equations of general relativity. Instead, Hoyle preferred to believe in a more sustainable vision of the cosmos. In his perpetual 'steady state' universe, matter and space were being continually created and destroyed and so could have existed for an unlimited time. Even so, clues were already amassing and by the 1960s Hoyle's static picture had to give way, given the weight of evidence that favoured the big bang.

timeline

AD 1927
Friedmann and Lemaître devise big bang theory

1929
Hubble detects the expansion of the universe

The expanding universe Three critical observations underpin the success of the big bang model. The first is Edwin Hubble's observation in the 1920s that most galaxies are moving away from our own. Looked at from afar, all galaxies tend to fly apart from one another as if the fabric of space–time is expanding and stretching, following Hubble's law. One consequence of the stretching is that light takes slightly longer to reach us when travelling across an expanding universe than one where distances are fixed. This effect is recorded as a shift in the frequency of the light, called the 'redshift' because the received light appears redder than it was when it left the distant star or galaxy. Redshifts can be used to infer astronomical distances.

> **'Tune your television to any channel it doesn't receive, and about 1% of the dancing static you see is accounted for by this ancient remnant of the big bang. The next time you complain that there is nothing on, remember that you can always watch the birth of the universe.'**
>
> Bill Bryson, 2005

Light elements Going back in time to the first hours of the newborn universe, just after the big bang, everything was packed close together in a seething superheated cauldron. Within the first second, the universe was so hot and dense that not even atoms were stable. As it grew and cooled a particle soup emerged first, stocked with quarks, gluons and other fundamental particles (see page 144). After just a minute the quarks stuck together to form protons and neutrons. Then, within the first three minutes, cosmic chemistry mixed the protons and neutrons, according to their relative numbers, into atomic nuclei. This is when elements other than hydrogen were first formed by nuclear fusion. Once the universe cooled below the fusion limit, no elements heavier than beryllium could be made. So the universe initially was awash with the nuclei of hydrogen, helium and traces of deuterium (heavy hydrogen), lithium and beryllium created in the big bang itself.

In the 1940s Ralph Alpher and George Gamow predicted the proportions of light elements produced in the big bang, and this basic picture has been confirmed by even the most recent measurements in slow-burning stars and primitive gas clouds in our Milky Way.

948
e cosmic microwave
ckground is predicted
bang nucleosynthesis is
culated by Alpher and Gamow

1949
Hoyle coins the term 'big bang'

1965
Penzias and Wilson detect the cosmic microwave background

1992
COBE satellite measures cosmic microwave background patches

Big bang timeline

13.7 billion years [after the big bang] Now (temperature, T = 2.726 K)

200 million years 'Reionization': first stars heat and ionize hydrogen gas (T = 50 K)

380 thousand years 'Recombination': hydrogen gas cools down to form molecules (T = 3000 K)

10 thousand years End of the radiation-dominated era (T = 12,000 K)

1000 seconds Decay of lone neutrons (T = 500 million K)

180 seconds 'Nucleosynthesis': formation of helium and other elements from hydrogen (T = 1 billion K)

10 seconds Annihilation of electron–positron pairs (T = 5 billion K)

1 second Decoupling of neutrinos (T ~ 10 billion K)

100 microseconds Annihilation of pions (T ~ 1 trillion K)

50 microseconds 'QCD phase transition': quarks bound into neutrons and protons (T = 2 trillion K)

10 picoseconds 'Electroweak phase transition': electromagnetic and weak force become different (T ~ 1–2 quadrillion K)

Before this time the temperatures were so high that our knowledge of physics is uncertain.

Time

Big bang

Microwave glow Another pillar supporting the big bang is the discovery in 1965 of the faint echo of the big bang itself. Arno Penzias and Robert Wilson were working on a radio receiver at Bell Labs in New Jersey when they were puzzled by a weak noise signal they could not get rid of. It seemed there was an extra source of microwaves coming from all over the sky, equivalent to something a few degrees in temperature.

After talking to astrophysicist Robert Dicke at nearby Princeton University, they realized that their signal matched predictions of the big bang afterglow. They had stumbled upon the cosmic microwave background radiation, a sea of photons left over from the very young hot universe. Dicke, who had built a similar radio antenna to look for the background radiation, was a little less jubilant: 'Boys, we've been scooped', he quipped.

In big bang theory, the existence of the microwave background had been predicted in 1948 by George Gamow, Ralph Alpher and Robert Hermann. Although nuclei were synthesized within the first three minutes, atoms were not formed for 400,000 years. Eventually, negatively charged electrons paired with positively charged nuclei to make atoms of hydrogen and light elements. The removal of charged particles, which scatter and block the path of light, cleared the fog and made the universe transparent. From then onwards, light could travel freely across the universe, allowing us to see back that far.

Although the young universe fog was originally hot (some 3000 kelvins), the expansion of the universe has redshifted the glow from it so that we see it today at a temperature of less than 3 K (three degrees above absolute zero). This is what Penzias and Wilson spotted. So with these three major foundations so far intact, big bang theory is widely accepted by most astrophysicists. A handful still pursue the steady state model that attracted Fred Hoyle, but it is difficult to explain all these observations in any other model.

‘There is a coherent plan in the universe, though I don’t know what it’s a plan for.’

Fred Hoyle, 1915–2001

Fate and past What happened before the big bang? Because space–time was created in it, this is not really a very meaningful question to ask – a bit like ‘where does the earth begin?’ or ‘what is north of the north pole on Earth?’. However, mathematical physicists do ponder the triggering of the big bang in multi-dimensional space (often 11 dimensions) through the mathematics of M-theory and string theory. These look at the physics and energies of strings, and membranes in these multi-dimensions and incorporate ideas of particle physics and quantum mechanics to try to trigger such an event. With parallels to quantum physics ideas, some cosmologists also discuss the existence of parallel universes.

In the big bang model, unlike the steady state model, the universe evolves. The cosmos’s fate is dictated largely by the balance between the amount of matter pulling it together through gravity and other physical forces that pull it apart, including the expansion. If gravity wins, then the universe’s expansion could one day stall and it could start to fall back in on itself, ending in a rewind of the big bang, known as the big crunch. Universes could follow many of these birth–death cycles. Alternatively, if the expansion and other repelling forces (such as dark energy) win, they will eventually pull all the stars and galaxies and planets apart and our universe could end up a dark desert of black holes and particles, a ‘big chill’. Lastly there is the ‘Goldilocks universe’, where the attractive and repellent forces balance and the universe continues to expand forever but gradually slows. It is this ending that modern cosmology is pointing to as being most likely. Our universe is just right.

the condensed idea
The ultimate explosion

46 Cosmic inflation

Why does the universe look the same in all directions? And why, when parallel light rays traverse space, do they remain parallel so we see separate stars? We think that the answer is inflation – the idea that the baby universe swelled up so fast in a split second that its wrinkles smoothed out and its subsequent expansion balanced gravity exactly.

The universe we live in is special. When we look out into it we see clear arrays of stars and distant galaxies without distortion. It could so easily be otherwise. Einstein's general relativity theory describes gravity as a warped sheet of space and time upon which light rays wend their way along curved paths (see page 164). So, potentially, light rays could become scrambled, and the universe we look out onto could appear distorted like reflections in a hall of mirrors. But overall, apart from the odd deviation as they skirt a galaxy, light rays tend to travel more or less in straight lines right across the universe. Our perspective remains clear all the way to the visible edge.

> ‘It is said that there's no such thing as a free lunch. But the universe is the ultimate free lunch.’
>
> **Alan Guth, b.1947**

Flatness Although relativity theory thinks of space–time as being a curved surface, astronomers sometimes describe the universe as ‘flat’, meaning that parallel light rays remain parallel no matter how far they travel through space, just as they would do if travelling along a flat plain. Space–time can be pictured as a rubber sheet, where heavy objects weigh down the sheet and rest in dips in it, representing

timeline

AD1981 Guth proposes inflation

1992 The COsmic Background Explorer (COBE) satellite detects hot and cold patches and measures their temperatures

Geometry of the universe

From the latest observations of the microwave background, such as those of the Wilkinson Microwave Anisotropy Probe (WMAP) satellite in 2003 and 2006, physicists have been able to measure the shape of space–time right across the universe. By comparing the sizes of hot and cold patches in the microwave sky with the lengths predicted for them by big bang theory, they show that the universe is 'flat'. Even over a journey across the entire universe lasting billions of years, light beams that set out parallel will remain parallel.

gravity. In reality, space–time has more dimensions (at least four: three of space and one of time) but it is hard to imagine those. The fabric is also continually expanding, following the big bang explosion. The universe's geometry is such that the sheet remains mostly flat, like a table top, give or take some small dips and lumps here or there due to the patterns of matter. So light's path across the universe is relatively unaffected, bar the odd detour around a massive body.

If there was too much matter, then everything would weigh the sheet down and it would eventually fold in on itself, reversing the expansion. In this scenario initially parallel light rays would eventually converge and meet at a point. If there were too little matter weighing it down, then the space time sheet would stretch and pull itself apart. Parallel light rays would diverge as they crossed it. However, our real universe seems to be somewhere in the middle, with just enough matter to hold the universe's fabric together while expanding steadily. So the universe appears to be precisely poised (see box).

2003

The Wilkinson Microwave Anisotropy Probe (WMAP) maps cosmic microwave background radiation

microwave background

One observation that encompasses all these problems is that of the cosmic microwave background radiation. This background marks the afterglow of the big bang fireball, redshifted now to a temperature of 2.73 K. It is precisely 2.73 K all over the sky, with hot and cold patches differing from this by as little as 1 part in 100,000. To this day this temperature measurement remains the most accurate one made for any body at a single temperature. This uniformity is surprising because when the universe was very young, distant regions of the universe could not communicate even at light speed. So it is puzzling that they nevertheless have exactly the same temperature. The tiny variations in temperature are the fossil imprints of the quantum fluctuations in the young universe.

Sameness Another feature of the universe is that it looks roughly the same in all directions. The galaxies do not concentrate in one spot, they are littered in all directions. This might not seem that surprising at first, but it is unexpected. The puzzle is that the universe is so big that its opposite edges should not be able to communicate even at the speed of light. Having only existed for 14 billion years, the universe is more than 14 billion light years across in size. So light, even though it is travelling at the fastest speed attainable by any transmitted signal, has not had time to get from one side of the universe to the other. So how does one side of the universe know what the other side should look like? This is the 'horizon problem', where the 'horizon' is the furthest distance that light has travelled since the birth of the universe, marking an illuminated sphere. So there are regions of space that we cannot and will never see, because light from there has not had time to travel to us yet.

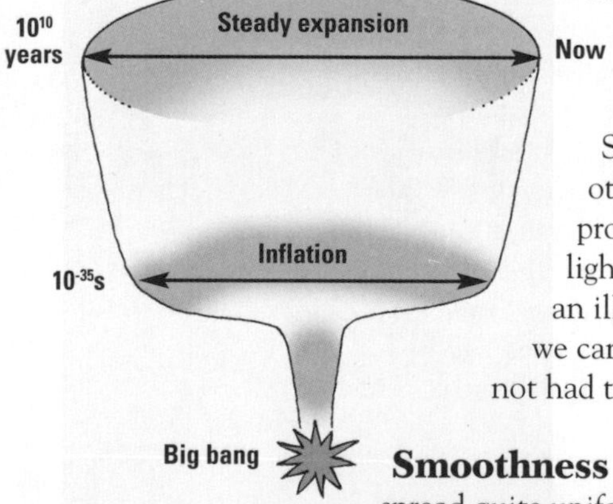

10^{10} years
Steady expansion
Now
Inflation
10^{-35}s
Big bang

Smoothness The universe is also quite smooth. Galaxies are spread quite uniformly across the sky. If you squint, they form a uniform glow rather than clumping in a few big patches. Again this need not have been the case. Galaxies have grown over time due to gravity. They started out as just a slightly overdense spot in the gas left over from the big bang.

That spot started to collapse due to gravity, forming stars and eventually building up a galaxy. The original overdense seeds of galaxies were set up by quantum effects, miniscule shifts in the energies of particles in the hot embryonic universe. But they could well have amplified to make large galaxy patches, like a cow's hide, unlike the widely scattered sea that we see. There are many molehills in the galaxy distribution rather than a few giant mountain ranges.

It is rather fantastic to realize that the laws of physics can describe how everything was created in a random quantum fluctuation out of nothing, and how over the course of 15 billion years, matter could organize in such complex ways that we have human beings sitting here, talking, doing things intentionally.

Alan Guth, b.1947

Growth spurt The flatness, horizon and smoothness problems of the universe can all be fixed with one idea: inflation. Inflation was developed as a solution in 1981 by American physicist, Alan Guth. The horizon problem, that the universe looks the same in all directions even though it is too large to know this, implies that the universe must at one time have been so small that light could communicate between all the regions in it. Because it is no longer like this, it must have then inflated quickly to the proportionately bigger universe we see now. But this period of inflation must have been extraordinarily rapid, much faster than the speed of light. The rapid expansion, doubling in size and doubling again and again in a split second, smeared out the slight density variations imprinted by quantum fluctuations, just like a printed pattern on an inflated balloon becomes fainter. So the universe became smooth. The inflationary process also fixed up the subsequent balance between gravity and the final expansion, proceeding at a much more leisurely pace thereafter. Inflation happened almost immediately after the big bang fireball (10^{-35} seconds after).

Inflation has not yet been proven and its ultimate cause is not well understood – there are as many models as theorists – but understanding it is a goal of the next generation of cosmology experiments, including the production of more detailed maps of the cosmic microwave background radiation and its polarization.

the condensed idea
Cosmic growth spurt

47 Dark matter

Ninety percent of the matter in the universe does not glow but is dark. Dark matter is detectable by its gravitational effect but hardly interacts with light waves or matter. Scientists think it may be in the form of MACHOs, failed stars and gaseous planets, or WIMPs, exotic subatomic particles – the hunt for dark matter is the wild frontier of physics.

Dark matter sounds exotic, and it may be, but its definition is quite down to Earth. Most of the things we see in the universe glow because they emit or reflect light. Stars twinkle by pumping out photons, and the planets shine by reflecting light from the Sun. Without that light, we simply would not see them. When the Moon passes into the Earth's shadow it is dark; when stars burn out they leave husks too faint to see; even a planet as big as Jupiter would be invisible if it was set free to wander far from the Sun. So it is perhaps not a big surprise that much of the stuff in the universe does not glow. It is dark matter.

Dark side Although we cannot see dark matter directly, we can detect its mass through its gravitational pull on other astronomical objects and also light rays. If we did not know the moon was there, we could still infer its presence because its gravity would tug and shift the orbit of the Earth slightly. We have even used the gravity-induced wobble applied to a parent star to discover planets around distant stars.

In the 1930s, Swiss astronomer Fritz Zwicky realized that a nearby giant cluster of galaxies was behaving in a way that implied its mass was much greater than the weight of all the stars in all the galaxies within it. He

timeline

AD1933
Zwicky measures dark matter in the Coma cluster

Energy budget

Today we know that only about 4% of the universe's matter is made up of baryons (normal matter comprising protons and neutrons). Another 23% is exotic dark matter. We do know that this isn't made up of baryons. It is harder to say what it is made from, but it could be particles such as WIMPs. The rest of the universe's energy budget consists of another thing entirely, dark energy.

inferred that some unknown dark matter accounted for 400 times as much material as luminous matter, glowing stars and hot gas, across the entire cluster. The sheer amount of dark matter was a big surprise, implying that most of the universe was not in the form of stars and gas but something else. So what is this dark stuff? And where does it hide?

Mass is also missing from individual spiral galaxies. Gas in the outer regions rotates faster than it should if the galaxy was only as heavy as the combined mass of stars within it. So such galaxies are more massive than expected by looking at the light alone. Again, the extra dark matter needs to be hundreds of times more abundant than the visible stars and gas. Dark matter is not only spread throughout galaxies but its mass is so great it dominates the motions of every star within them. Dark matter even extends beyond the stars, filling a spherical 'halo' or bubble around every flattened spiral galaxy disk.

Weight gain Astronomers have now mapped dark matter not only in individual galaxies but also in clusters of galaxies, containing thousands of galaxies bound together by mutual gravity, and superclusters of galaxies,

1975
Vera Rubin shows that galaxy rotation is affected by dark matter

1998
Neutrinos inferred to have a small mass

2000
MACHOs detected in the Milky Way

chains of clusters in a vast web that stretches across all of space. Dark matter features wherever there is gravity at work, on every scale. If we add up all the dark matter, we find that there is a thousand times more dark stuff as luminous matter.

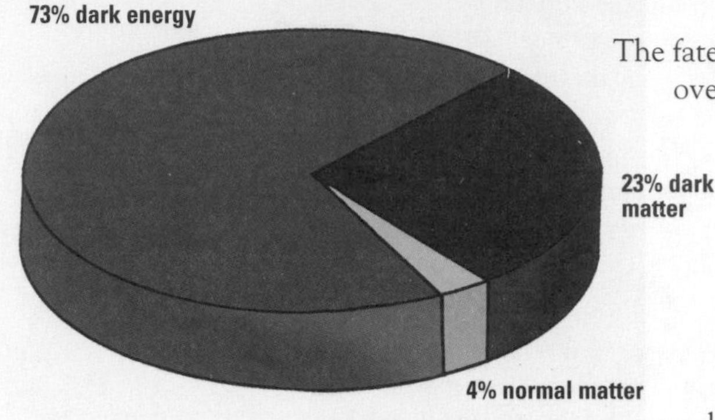

The fate of the entire universe depends on its overall weight. Gravity's attraction counterbalances the expansion of the universe following the big bang explosion. There are three possible outcomes. Either the universe is so heavy that gravity wins and the universe eventually collapses back in on itself (a closed universe ending in a big crunch), or there is too little mass and it expands forever (an open universe), or the universe is precisely balanced and the expansion gradually slows by gravity, but over such a long time that it never ceases. The latter seems the best case for our universe, it has precisely the right amount of matter to slow but never halt the expansion.

WIMPs and MACHOs What might dark matter be made of? First, it could be dark gas clouds, dim stars or unlit planets. These are called MACHOs, or MAssive Compact Halo Objects. Alternatively the dark matter could be new kinds of subatomic particles, called WIMPs, short for Weakly Interacting Massive Particles, which would have virtually no effect on other matter or light.

Astronomers have found MACHOs roaming within our own Galaxy. Because MACHOs are large, akin to the planet Jupiter, they can be spotted individually by their gravitational effect. If a large gas planet or failed star passes in front of a background star, its gravity bends the starlight around it. The bending focuses the light when the MACHO is right in front of the star, so the star appears much brighter for a moment as it passes. This is called 'gravitational lensing'.

In terms of relativity theory, the MACHO planet distorts space–time, like a heavy ball depressing a rubber sheet, which curves the light's wavefront around it (see page 164). Astronomers have looked for this brightening of

> **"The universe is made mostly of dark matter and dark energy, and we don't know what either of them is."**
>
> **Saul Perlmutter, 1999**

stars by the passage of a foreground MACHO against millions of stars in the background. They have found a few such flare ups, but too few to explain all the missing mass of the Milky Way.

MACHOs are made of normal matter, or baryons, built of protons, neutrons and electrons. The tightest limit on the amount of baryons in the universe is given by tracking the heavy hydrogen isotope, deuterium. Deuterium was only produced in the big bang itself and is not formed by stars afterwards, although it can be burned within them. So, by measuring the amount of deuterium in pristine gas clouds in space, astronomers can estimate the total number of protons and neutrons that were made in the big bang, because the mechanism for making deuterium is precisely known. This turns out to be just a few percent of the mass of the entire universe. So the rest of the universe must be in some entirely different form, such as WIMPs.

The search for WIMPS is now the focus of attention. Because they are weakly interacting these particles are intrinsically difficult to detect. One candidate is the neutrino. In the last decade physicists have measured its mass and found it to be very small but not zero. Neutrinos make up some of the universe's mass, but again not all. So there is still room for other more exotic particles out there waiting to be detected, some new to physics such as axions and photinos. Understanding dark matter may yet light up the world of physics.

the condensed idea
Dark side of the universe

48 Cosmological constant

Einstein called adding his cosmological constant into the equations of general relativity his biggest blunder. The term allowed for the speeding up or slowing down of the rate of expansion of the universe to compensate gravity. Einstein did not need this number and abandoned it. However, new evidence in the 1990s required that it be reintroduced. Astronomers found that mysterious dark energy is causing the expansion of the universe to speed up, leading to the rewriting of modern cosmology.

Albert Einstein thought we lived in a steady state universe rather than one with a big bang. Trying to write down the equations for it, he ran into a problem. If you just had gravity, then everything in the universe would ultimately collapse into a point, perhaps a black hole. Obviously the real universe wasn't like that and appeared stable. So Einstein added another term to his theory to counterbalance gravity, a sort of repulsive 'anti-gravity' term. He introduced this purely to make the equations look right, not because he knew of such a force. But this formulation was immediately problematic.

If there was a counterforce to gravity then just as untrammelled gravity could cause collapse, then an anti-gravity force could just as easily amplify to tear apart regions of the universe that were not held together by gravity's glue. Rather than allow such shredding of the universe, Einstein preferred to ignore his second repulsive term and admitted he had made a

timeline

AD1915
Einstein publishes the general theory of relativity

1929
Hubble shows space is expanding and Einstein abandons his constant

mistake in introducing it. Other physicists also preferred to exclude it, relegating it to history. Or so they thought. The term was not forgotten – it was preserved in the relativity equations but its value, the cosmological constant, was set to zero to dismiss it.

Accelerating universe In the 1990s, two groups of astronomers were mapping supernovae in distant galaxies to measure the geometry of space and found that distant supernovae appeared fainter than they should be. Supernovae, the brilliant explosions of dying stars, come in many types. Type Ia supernovae have a predictable brightness and so are useful for inferring distances. Just like the Cepheid variable stars that were used to measure the distances to galaxies to establish Hubble's law, the intrinsic brightness of Type Ia supernovae can be worked out from their light spectra so that it is possible to say how far away they must be. This all worked fine for supernovae that were quite nearby, but the more distant supernovae were too faint. It was as if they were further away from us than they should be.

> **'For 70 years, we've been trying to measure the rate at which the universe slows down. We finally do it, and we find out it's speeding up.'**
>
> **Michael S. Turner, 2001**

As more and more distant supernovae were discovered, the pattern of the dimming with distance began to suggest that the expansion of the universe was not steady, as in Hubble's law, but was accelerating. This was a profound shock to the cosmology community, and one that is still being disentangled today.

The supernova results fitted well with Einstein's equations, but only once a negative term was included by raising the cosmological constant from zero to about 0.7. The supernova results, taken with other cosmological data, such as the cosmic microwave background radiation pattern, showed that a new repulsive force counteracting gravity was needed. But it was quite a weak force. It is still a puzzle today why it is so weak, as there is no particular reason why it did not adopt a much larger value and perhaps

1998

Supernova data indicates the need for the cosmological constant

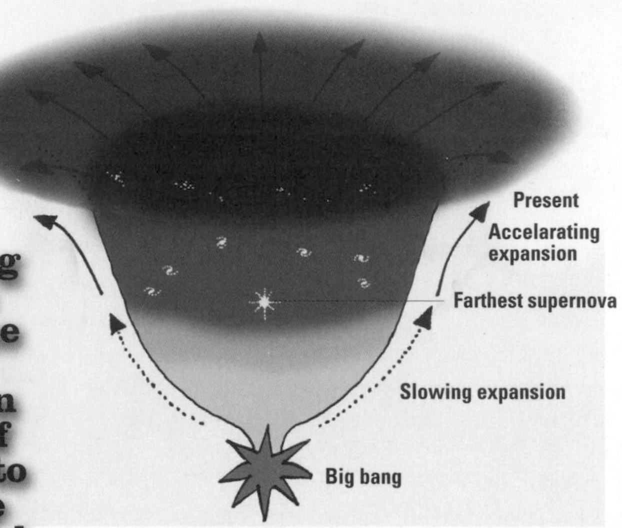

‘It [dark energy] seems to be something that is connected to space itself and unlike dark matter which gravitates this has an effect which is sort of the opposite, counter to gravity, it causes the universe to be repulsed by itself.’

Brian Schmidt, 2006

completely dominate space over gravity. Instead it is very close in strength to gravity so has a subtle effect on space time as we see it now. This negative energy term has been named ‘dark energy’.

Dark energy Dark energy’s origin is still elusive. All we know is that it is a form of energy associated with the vacuum of free space, causing a negative pressure in regions devoid of gravity-attracting matter. So it causes regions of empty space to inflate. We know its strength roughly from the supernova observations, but we do not know much more. We don’t know if it truly is a constant – whether it always takes the same value right across the universe and for all time (as do gravity and the speed of light) – or whether its value changes with time so it may have had a different value just after the big bang compared with now or in the future. In its more general form it has also been called ‘quintessence’ or the fifth force, encompassing all the possible ways its strength could change with time. But it is still not known how this elusive force manifests itself or how it arises within the physics of the big bang. It is a hot topic of study for physicists.

Nowadays we have a much better understanding of the geometry of the universe and what it is made up of. The discovery of dark energy has balanced the books of cosmology, making up the difference in the energy budget of the whole universe. So we now know that it is 4% normal baryonic matter, 23% exotic non-baryonic matter, and 73% dark energy.

‘It is to be emphasized, however, that a positive curvature of space is given by our results, even if the supplementary term [cosmological constant] is not introduced. That term is necessary only for the purpose of making possible a quasi-static distribution of matter.’

Albert Einstein, 1918

These numbers add up to about the right amount of stuff for the balanced ‘Goldilocks universe’, close to the critical mass where it is neither open nor closed.

The mysterious qualities of dark energy however mean that even knowing the total mass of the universe, its future behaviour is hard to predict because it depends on whether the influence of dark energy increases or not in the future. If the universe is accelerating then, at this point in time, dark energy is only just as significant as gravity in dominating the universe. But, at some point, the acceleration will pick up and the faster expansion will overtake gravity. So the universe’s fate may well be to expand forever, faster and faster. Some scary scenarios have been proposed – once gravity is outstripped, then tenuously held together massive structures will disconnect and fly apart, eventually even galaxies themselves will break up, then stars will be evaporated into a mist of atoms. Ultimately the negative pressure could strip atoms, leaving only a grim sea of subatomic particles.

Nevertheless, although cosmology’s jigsaw is fitting together now, and we have measured a lot of the numbers that describe the geometry of the universe, there are still some big unanswered questions. We just don’t know what 95% of the stuff in the universe is, nor what this new force of quintessence really is. So it is not yet time to sit back and rest on our laurels. The universe has kept its mystery.

the condensed idea
The fifth force

49 Fermi paradox

The detection of life elsewhere in the universe would be the greatest discovery of all time. Enrico Fermi wondered why, given the age and vastness of the universe, and the presence of billions of stars and planets that have existed for billions of years, we have not yet been contacted by any other alien civilizations. This was his paradox.

Chatting with his colleagues over lunch in 1950, physics professor Enrico Fermi supposedly asked 'Where are they?' Our own Galaxy contains billions of stars and there are billions of galaxies in the universe, so that is trillions of stars. If just a fraction of those anchored planets, that's a lot of planets. If a fraction of those planets sheltered life, then there should be millions of civilizations out there. So why haven't we seen them? Why haven't they got in touch with us?

Drake equation In 1961, Frank Drake wrote down an equation for the probability of a contactable alien civilization living on another planet in the Milky Way. This is known as the Drake equation. It tells us that there is a chance that we may coexist with another civilization but the probability is still quite uncertain. Carl Sagan once suggested that as many as a million alien civilizations could populate the Milky Way, but he later revised this down and others since have estimated that the value is just one, namely humans. More than half a century after Fermi asked the question, we have still heard nothing. Despite our communication systems, no one has called. The more we explore our local neighbourhood, the lonelier it seems. No concrete signs of any life, not even the simplest

AD 1950
Fermi questions the absence of alien contact

1961
Drake devises his equation

bacteria, have been found on the Moon, Mars, asteroids, the outer solar system planets and moons. There are no signs of interference in the light from stars that could indicate giant orbiting machines harvesting energy from them. And it is not because no one has been looking. Given the stakes there is great attention given to searching for extraterrestrial intelligence.

> **Who are we? We find that we live on an insignificant planet of a humdrum star lost in a galaxy tucked away in some forgotten corner of a universe in which there are far more galaxies than people.**
>
> **Werner von Braun, 1960**

Search for life So, how would you go about looking for signs of life? The first way is to start looking for microbes within our solar system. Scientists have scrutinized rocks from the moon but they are inanimate basalt. Meteorites from Mars have been suggested to host the remnants of bacteria, but it is still not proven that the ovoid bubbles in those rocks hosted alien life and were not contaminated after having fallen to Earth or produced by natural geological processes. Even without rock samples, cameras on spacecraft and landers have scoured the surfaces of Mars, asteroids and now even a moon in the outer solar system – Titan, orbiting Saturn.

But the Martian surface is a dry desert of volcanic sand and rocks, not unlike the Atacama desert in Chile. Titan's surface is damp, drenched in liquid methane, but so far devoid of life. One of Jupiter's moons, Europa, has been touted as a popular target for future searches for life in the solar system, as it may host seas of liquid water beneath its frozen surface. Space scientists are planning a mission there that will drill through the ice crust and look below. Other moons in the outer solar system have been found to be quite geologically active, releasing heat as they are squeezed and pulled by the gravitational torques of their orbits around the giant gas planets. So liquid water may not be so rare a commodity in the outer solar system, raising expectations that one day life may be found. Spacecraft that go into this region are extensively sterilized to make sure that we do not contaminate them with foreign microbes from Earth.

1996

Antarctic meteorites hint at primitive life existing on Mars

Drake equation

$N = N_* \times f_p \times n_e \times f_l \times f_i \times f_c \times f_L$

where:

N is the number of civilizations in the Milky Way Galaxy whose electromagnetic emissions are detectable

N_* is the number of stars in the Galaxy

f_p is the fraction of those stars with planetary systems

n_e is the number of planets, per solar system, with an environment suitable for life

f_l is the fraction of suitable planets on which life actually appears

f_i is the fraction of life-bearing planets on which intelligent life emerges

f_c is the fraction of civilizations that develop a technology that releases detectable signs of their existence into space

f_L is the fraction of a planetary lifetime such civilizations release detectable signals into space (for earth this fraction is so far very small).

But microbes are not going to call home. What about more sophisticated animals or plants? Now that individual planets are being detected around distant stars, astronomers are planning on dissecting the light from them to hunt for chemistry that could support or indicate life. Spectral hints of ozone or chlorophyll might be picked up, but these will need precise observations like those possible with the next generation of space missions such as NASA's Terrestrial Planet Finder. These missions might find us a sister Earth one day, but if they did, would it be populated with humans, fish or dinosaurs, or just contain empty lifeless continents and seas?

Contact Life on other planets, even Earth-like ones, might have evolved differently to that on Earth. So it is not certain that aliens there would be able to communicate with us on Earth. Since radio and television began broadcasting, their signals have been spreading away from Earth, travelling outwards at the speed of light. So any TV fan on Alpha Centauri (4 light years away) would be watching the Earth channels from 4 years ago, perhaps enjoying repeats of the film *Contact*. Black and white movies would be reaching the star Arcturus, and Charlie Chaplin could be starring at Aldebaran. So, Earth is giving off plenty of signals, if you have an antenna to pick them up. Wouldn't other advanced civilizations do the same? Radio astronomers are scouring nearby stars for signs of unnatural signals.

> **‘Our sun is one of 100 billion stars in our galaxy. Our galaxy is one of billions of galaxies populating the universe. It would be the height of presumption to think that we are the only living things in that enormous immensity.’**
>
> **Carl Sagan,** 1980

The radio spectrum is vast, so they are focusing on frequencies near key natural energy transitions, such as those of hydrogen, which should be the same anywhere in the universe. They are looking for transmissions that are regular or structured but not made by any known astronomical objects. In 1967, graduate student Jocelyn Bell got a fright in Cambridge, England, when she discovered regular pulses of radio waves coming from a star. Some thought this was indeed an alien Morse code, but in fact it was a new type of spinning neutron star now called a pulsar. Because this process of scanning thousands of stars takes a long time, a special programme has been started in the USA called SETI (Search for ExtraTerrestrial Intelligence). Despite analysing years of data, no odd signals have yet been picked up. Other radio telescopes search occasionally, but these too have seen nothing that does not have a more mundane origin.

Out to lunch So, given that we can think of many ways to communicate and detect signs of life, why might any civilizations not be returning our calls or sending their own? Why is Fermi's paradox still true? There are many ideas. Perhaps life only exists for a very short time in an advanced state where communication is possible.Why might this be? Perhaps intelligent life always wipes itself out quickly. Perhaps it is self-destructive and does not survive long, so the chances of being able to communicate and having someone nearby to communicate to are very low indeed. Or there are more paranoid scenarios. Perhaps aliens simply do not want to contact us and we are deliberately isolated. Or, perhaps they are just too busy and haven't got around to it yet.

the condensed idea
Is there anybody out there?

50 Anthropic principle

The anthropic principle states that the universe is as it is because if it were different we would not be here to observe it. It is one explanation for why every parameter in physics takes the value that it does, from the size of the nuclear forces to dark energy and the mass of the electron. If any one of those varied even slightly then the universe would be uninhabitable.

If the strong nuclear force was a little different then protons and neutrons would not stick together to make nuclei and atoms could not form. Chemistry would not exist. Carbon would not exist and so biology and humans would not exist. If we did not exist, who would 'observe' the universe and prevent it from existing only as a quantum soup of probability?

Equally, even if atoms existed and the universe had evolved to make all the structures we know today, then if dark energy were a little stronger galaxies and stars would already be being pulled apart. So, tiny changes in the values of the physical constants, in the sizes of forces or masses of particles, can have catastrophic implications. Put another way, the universe appears to be fine tuned. The forces are all 'just right' for humanity to have evolved now. Is it a chance happening that we are living in a universe that is 14 billion years old, where dark energy and gravity balance each other out and the subatomic particles take the forms they do?

timeline

AD 1904
Alfred Wallace discusses man's place in the universe

1957
Robert Dicke writes that the universe is constrained by biological factors

Just so Rather than feel that humanity is particularly special and the entire universe was put in place just for us, perhaps a rather arrogant assumption, the anthropic principle explains that it is no surprise. If any of the forces were slightly different, then we simply would not be here to witness it. Just as the fact that there are many planets but as far as we know only one that has the right conditions for life, the universe could have been made in many ways but it is only in this one that we would come to exist. Equally, if my parents had never met, if the combustion engine had not been invented when it was and my father had not been able to travel north to meet my mother, then I would not be here. That does not mean that the entire universe evolved thus just so that I could exist. But the fact that I exist ultimately required, amongst other things, that the engine was invented beforehand, and so narrows the range of universes that I might be found in.

‘The observed values of all physical and cosmological quantities are not equally probable but they take on values restricted by the requirement that there exist sites where carbon-based life can evolve and . . . that the Universe be old enough for it to have already done so.’

John Barrow and Frank Tipler, 1986

The anthropic principle was used as an argument in physics and cosmology by Robert Dicke and Brandon Carter, although its argument is familiar to philosophers. One formulation, the weak anthropic principle, states that we would not be here if the parameters were different, so the fact that we exist restricts the properties of inhabitable physical universes that we could find ourselves in. Another stronger version emphasizes the importance of our own existence, such that life is a necessary outcome for the universe coming into being. For example, observers are needed to make a quantum universe concrete by observing it. John Barrow and Frank Tipler also suggested yet another version, whereby information processing is a fundamental purpose of the universe and so its existence must produce creatures able to process information.

1973

Brandon Carter discusses the anthropic principle

Anthropic bubbles

We can avoid the anthropic dilemma if many parallel, or bubble, universes accompany the one we live in. Each bubble universe can take on slightly different parameters of physics. These govern how each universe evolves and whether a given one provides a nice niche in which life can form. As far as we know, life is fussy and so will choose only a few universes. But since there are so many bubble universes, this is a possibility and so our existence is not so improbable.

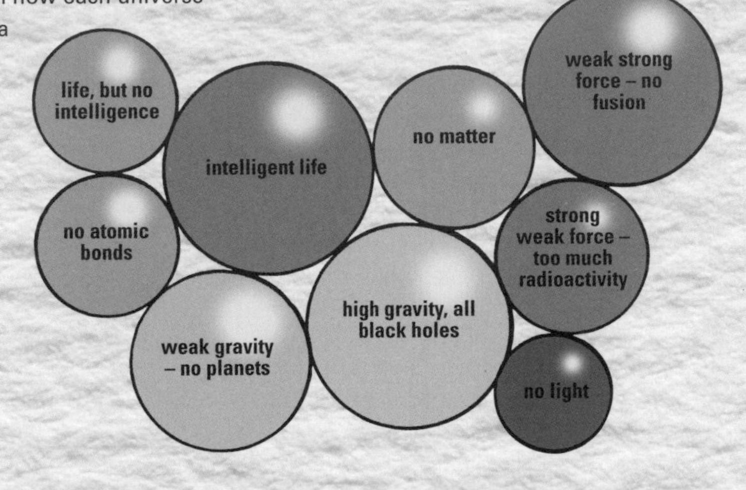

Many worlds To produce humans, you need the universe to be old, so that there's enough time for carbon to be made in earlier generations of stars, and the strong and weak nuclear forces must be 'just so' to allow nuclear physics and chemistry. Gravity and dark energy must also be in balance to make stars rather than rip apart the universe. Further, stars need be long lived to let planets form, and large enough so that we can find ourselves on a nice suburban temperate planet that has water, nitrogen, oxygen and all the other molecules needed to seed life.

Because physicists can imagine universes where these quantities are different, some have suggested that those universes can be created just as readily as one like our own. They may exist as parallel universes, or multi-verses, such that we only exist in one realization.

The idea of parallel universes fits in with the anthropic principle in allowing other universes to also exist where we cannot. These may exist in multiple dimensions and are split off along the same lines as quantum theory requires for observations to trigger outcomes (see page 115).

On the other hand The anthropic principle has its critics. Some think it is a truism – it is like this because it is like this – and is not telling us much that's new. Others are unhappy that we have just this one special universe to test, and prefer to search the mathematics for ways of automatically tuning our universe to fall out of the equations simply because of the physics. The multiverse idea comes close to this by allowing an infinite number of alternatives. Yet other theorists, including string theorists and proponents of M-theory, are trying to go beyond the big bang to fine tune the parameters. They look at the quantum sea that preceded the big bang as a sort of energy landscape and ask where a universe is most likely to end up if you let it roll and unfold. For instance, if you roll a ball down a ridged hill, then the ball is more likely to end up in some places than others, such as in valley floors. So in trying to minimize its energy, the universe may well seek out certain combinations of parameters quite naturally, irrespective of whether we are a product of it billions of years later.

‘In order to make an apple pie from scratch, you must first create the universe.’

Carl Sagan, 1980

Proponents of the anthropic principle and others, who pursue more mathematical means of ending up with the universe we know, disagree about how we got to be where we are and whether that is even an interesting question to ask. Once we get beyond the big bang and the observable universe, into the realms of parallel universes and pre-existing energy fields, we are really on philosophical ground. But whatever triggered the universe to appear in its current garb, we are lucky that it has turned out this way billions of years hence. It is understandable that it takes time to cook up the chemistry needed for life. But why we should be living here at a particular time in the universe's history when dark energy is relatively benign and balancing out gravity is more than lucky.

the condensed idea
The just so universe

Glossary

Acceleration The change in something's velocity in a given time.

Age of the universe *see* Universe

Atom The smallest unit of matter that can exist independently. Atoms contain a hard central nucleus made up of (positively charged) protons and (uncharged) neutrons surrounded by clouds of (negatively charged) electrons.

Black-body radiation Light glow emitted by a black object at a specific temperature, which has a characteristic spectrum.

Boson A particle with a symmetric wave function; two bosons can occupy the same quantum state (*see also* Fermion).

Cosmic microwave background radiation A faint microwave glow that fills the sky. It is the afterglow of the big bang that has since cooled and been redshifted to a temperature of 3 kelvins.

Diffraction The spreading out of waves when they pass a sharp edge, such as water waves entering a harbour through a gap in the wall.

Elasticity Elastic materials obey Hooke's law. They stretch by an amount that is proportional to the force applied.

Electricity The flow of electric charge. It has some voltage (energy), may cause a current (a flow) and can be slowed or blocked by resistance.

Energy A property of something that dictates its potential for change. It is conserved overall but can be exchanged between many different types.

Entanglement In quantum theory, the idea that particles that are related at one point in time carry away information with them thereafter and can be used for instantaneous signalling.

Entropy A measure of disorder. The more ordered something is, the lower its entropy.

Fermion A particle that follows Pauli's exclusion principle, where no two fermions can have the same quantum state (*see also* Boson).

Fields A means of transmitting a force at a distance. Electricity and magnetism are fields, as is gravity.

Force A lift, pull or push, causing the motion of something to change. Newton's 2nd law defines a force as being proportional to the acceleration it produces.

Frequency The rate at which wave crests pass some point.

Galaxy A group or cloud of millions of stars held together by gravity. Our own Milky Way is a spiral galaxy.

Gas A cloud of unbound atoms or molecules. Gases have no edges but may be confined by a container.

Gravity A fundamental force through which masses attract one another. Gravity is described by Einstein's theory of general relativity.

Inertia *see* Mass

Interference The combining of waves of different phases that may produce reinforcement (if in phase) or cancellation (if out of phase).

Isotope A chemical element existing in different forms, with the same number of protons but a different number of neutrons in its nucleus, so with different atomic masses.

Many-worlds hypothesis In quantum theory and cosmology, the idea that there are many parallel universes that branch off as events occur, and that we are at any time in one branch.

Mass A property that is equivalent to the number of atoms or amount of energy that something contains. Inertia is a similar idea that describes mass in terms of its resistance to movement, such that a heavy (more massive) object is harder to move.

Momentum The product of mass and velocity that expresses how hard it is to stop something once moving.

Nucleus The hard central core of the atom, made of protons and neutrons held together by the strong nuclear force.

Observer In quantum theory, an observer is someone who performs an experiment and measures the outcome.

Phase The relative shift between one wave and another measured in wavelength fractions. One whole wavelength shift is 360 degrees; if the relative shift is 180 degrees, the two waves are exactly out of phase (*see also* Interference).

Photon Light manifesting as a particle.

Pressure Defined as the force per unit area. The pressure of a gas is the force exerted by its atoms or molecules on the inside surface of its container.

Quanta The smallest sub-units of energy, as used in quantum theory.

Quark A fundamental particle, three of which combine to make up protons and neutrons. Forms of matter made of quarks are called hadrons.

Qubits Quantum bits. Similar to computer 'bits' but including quantum information.

Randomness A random outcome is determined only by chance. No particular outcomes are favoured.

Redshift The shift in wavelength of light from a receding object, due to the Doppler effect or cosmological expansion. In astronomy it is a way of measuring distances to far away stars and galaxies.

Reflection The reversal of a wave when it hits a surface, such as a light beam bouncing off a mirror.

Refraction The bending of waves, usually due to their slowing down as they pass through a medium, such as light through a prism.

Space–time metric Geometric space combined with time into one mathematical function in general relativity. It is often visualized as a rubber sheet.

Spectrum The sequence of electromagnetic waves, from radio waves through visible light to X-rays and gamma rays.

Strain The amount by which something extends when it is pulled, per unit length.

Stress Force per unit area, felt internally by a solid due to a load being applied to it.

Supernova The explosion of a star above a certain mass when it reaches the end of its life.

Turbulence When fluid flows become too fast they become unstable and turbulent, breaking down into swirls and eddies.

Universe All of space and time. By definition it includes everything, but some physicists talk of parallel universes separate from our own. Our universe is about 14 billion years old, determined from its rate of expansion and the ages of stars.

Vacuum A space that contains no atoms is a vacuum. None exists in nature – even outer space has a few atoms per cubic centimetre – but physicists come close in the laboratory.

Velocity Velocity is speed in a particular direction. It is the distance in that direction by which something moves in a given time.

Wave function In quantum theory, a mathematical function that describes all the characteristics of some particle or body, including the probability that it has certain properties or is in some location.

Wavefront The line tracing the peak of a wave.

Wavelength The distance from one wave crest to the next adjacent one.

Wave–particle duality Behaviour, particularly of light, that is sometimes wave-like and at other times like a particle.

Index

Quercus Publishing Plc
21 Bloomsbury Square
London
WC1A 2NS

First published in 2007

A catalogue record of this book is available from the British Library

ISBNs
Cloth case edition ISBN-10: 1 84724 007 0
ISBN-13: 978 1 84724 007 1

Printed case edition ISBN-10: 1 84724 148 4
ISBN-13: 978 1 84724 148 1

Printed and bound in China

10 9

Picture credits: p.46: iStockphotos

Edited by Keith Mansfield
Designed and illustrated by Patrick Nugent
Proofread by John Holmes
Index by Ingrid Lock